Supported by the National Fund for Academic Publication in Science and Technology
Supported by the National Key R&D Program of China (Grant No. 2016YFC0800200)

SOLUTIONS FOR BIOT'S POROELASTIC THEORY IN KEY ENGINEERING FIELDS

Theory and Applications

Yuanqiang Cai
Honglei Sun

ZHEJIANG UNIVERSITY PRESS
浙江大学出版社

ELSEVIER

Introduction

All structures made by human beings have to be placed on or in the soil. At the very beginning, these soil foundations are only subjected to static loads, which are loads that build up gradually over time or with negligible dynamic effects, also known as monotonic loads. In the 1930s, due to the rapid development of machine manufacturing and transportation industries, the dynamic interaction between structural foundations and the underlying soil behavior because of the action of cyclic loads started to receive considerable attention in a number of engineering fields. Cyclic loads are loads which exhibit a degree of regularity both in their magnitude and frequency. Stress reversals, rate effects and dynamic effects are the important factors that distinguish cyclic loads from static loads.

Practically speaking, no real cyclic loads exist in nature; however, many kinds of loads can be simplified into cyclic loads for the convenience of study, analysis and design. For example, the operation of a reciprocating or a rotary machine typically produces a cyclic load. The passing of a long train can be considered a cyclic load. Even cars running on a road on the same line can be simplified as a cyclic load. Cyclic loads acting on the structures and soil can produce elastic waves in the ground which will act on the surrounding foundations and soil. These actions can cause environmental and safety problems. As a result, it is very important to take a deep look into this area to advance the knowledge regarding the theory of vibrations, the principles of wave propagation, and numerical methods in finding appropriate solutions to problems of practical interest.

In the past few decades, many studies have been carried out on soil-structure interactions under cyclic loads. Most of them have treated the soil as an elastic or viscoelastic medium. However, there is underground water in what is considered soil medium, so the soil is actually a two-phase medium. Biot[1] pioneered the development of an elastodynamic theory for a fluid-filled elastic porous medium. Since its publication, Biot's theory has had wide applications in the geotechnical professions for analyzing wave propagation characteristics under cyclic loads. The aim of this book is to provide a tutorial and a state-of-the-art compilation of the advances in the applications of Biot's theory.

Contents

Chapter 1
Basic Equations and Governing Equations

This chapter introduces the basic equations, the governing equations and the boundary conditions of Biot's theory and its transformed forms, which are essential for solving engineering problems in a fully saturated poroelastic medium.

We use two kinds of coordinate systems in this book, a Cartesian coordinate system and a cylindrical coordinate system. The relationships between these coordinate systems are given in Table 1.1.1, in which angle θ ($0 \le \theta \le 2\pi$) is measured from the positive direction of the x-axis to that of the y-axis. In these coordinates, we use (u_x, u_y, u_z), (u_r, u_θ, u_z) to denote the displacements at a point in the solid in Cartesian and cylindrical coordinate systems, respectively. The footnotes demonstrate along which direction the displacement occurs. We can also use them in the matrix form as $\{u\} = [u_x, u_y, u_z]^{\mathrm{T}}$ or $[u_r, u_\theta, u_z]^{\mathrm{T}}$, respectively, where the superscript T stands for transpose.

Table 1.1.1 Direction cosines between coordinate axes in Cartesian and cylindrical coordinates

Angel	x	y	z
r	$\cos\theta$	$\sin\theta$	0
θ	$-\sin\theta$	$\cos\theta$	0
z	0	0	1

We write the stresses and strains in the Cartesian and cylindrical coordinate systems, respectively, as

$$\begin{bmatrix} \sigma_{xx} & \tau_{xy} & \tau_{zx} \\ \tau_{xy} & \sigma_{yy} & \tau_{yz} \\ \tau_{zx} & \tau_{yz} & \sigma_{zz} \end{bmatrix} \quad \text{and} \quad \begin{bmatrix} \varepsilon_{xx} & \frac{1}{2}\gamma_{xy} & \frac{1}{2}\gamma_{zx} \\ \frac{1}{2}\gamma_{xy} & \varepsilon_{yy} & \frac{1}{2}\gamma_{yz} \\ \frac{1}{2}\gamma_{zx} & \frac{1}{2}\gamma_{yz} & \varepsilon_{zz} \end{bmatrix}$$

$$\begin{bmatrix} \sigma_{rr} & \tau_{r\theta} & \tau_{zr} \\ \tau_{r\theta} & \sigma_{\theta\theta} & \tau_{\theta z} \\ \tau_{zr} & \tau_{\theta z} & \sigma_{zz} \end{bmatrix} \quad \text{and} \quad \begin{bmatrix} \varepsilon_{rr} & \frac{1}{2}\gamma_{r\theta} & \frac{1}{2}\gamma_{zr} \\ \frac{1}{2}\gamma_{r\theta} & \varepsilon_{\theta\theta} & \frac{1}{2}\gamma_{\theta z} \\ \frac{1}{2}\gamma_{zr} & \frac{1}{2}\gamma_{\theta z} & \varepsilon_{zz} \end{bmatrix}$$

where σ_{ij} is the stress tensor, τ_{ij} is the shear stress, γ_{ij} is the shear strain, and ε_{ij} is the strain tensor.

The transformation rules between the stresses and strains in different coordinates are

$$\sigma_{p'q'}=l_{p'i}\,l_{q'j}\,\sigma_{ij} \tag{1.1.1}$$

$$\varepsilon_{p'q'}=l_{p'i}\,l_{q'j}\,\varepsilon_{ij} \tag{1.1.2}$$

where σ_{ij} denotes the stresses in Cartesian coordinates, $\sigma_{p'q'}$ stands for the stresses in a new Cartesian coordinate system after rotation, and $l_{p'i}$ and $l_{q'j}$ represent the direction cosines between two coordinate axes.

Combining Eq. (1.1.1) with Table 1.1.1, the relationships between the stresses in cylindrical coordinates and Cartesian coordinates can be easily derived:

$$\begin{aligned}
&\sigma_r=\sigma_x\cos^2\theta+\sigma_y\sin^2\theta+2\tau_{xy}\sin\theta\cos\theta\\
&\sigma_\theta=\sigma_x\sin^2\theta+\sigma_y\cos^2\theta-2\tau_{xy}\sin\theta\cos\theta\\
&\tau_{r\theta}=\left(\sigma_y-\sigma_x\right)\sin\theta\cos\theta+\tau_{xy}\left(\cos^2\theta-\sin^2\theta\right)\\
&\tau_{zr}=\tau_{zx}\cos\theta+\tau_{yz}\sin\theta\\
&\tau_{\theta z}=-\tau_{zx}\sin\theta+\tau_{yz}\cos\theta\\
&\sigma_z=\sigma_z
\end{aligned} \tag{1.1.3}$$

1.1 Basic Equations

The basic equations for a linear elastic medium are geometric equations (strain-displacement relations), equations of motion and constitutive equations (stress-strain relations).

1.1.1 Geometric Equations

The geometric equations in Cartesian coordinates are given as

$$\begin{aligned}
&\varepsilon_x=\frac{\partial u_x}{\partial x},\quad \gamma_{yz}=\frac{\partial u_y}{\partial z}+\frac{\partial u_z}{\partial y}\\
&\varepsilon_y=\frac{\partial u_y}{\partial y},\quad \gamma_{zx}=\frac{\partial u_z}{\partial x}+\frac{\partial u_x}{\partial z}\\
&\varepsilon_z=\frac{\partial u_z}{\partial z},\quad \gamma_{xy}=\frac{\partial u_x}{\partial y}+\frac{\partial u_y}{\partial x}
\end{aligned} \tag{1.1.4}$$

In cylindrical coordinates, they become

$$
\begin{aligned}
&\varepsilon_r = \frac{\partial u_r}{\partial r}, \quad \gamma_{\theta z} = \frac{\partial u_\theta}{\partial z} + \frac{\partial u_z}{r\partial \theta} \\
&\varepsilon_\theta = \frac{1}{r}\frac{\partial u_\theta}{\partial \theta} + \frac{u_r}{r}, \quad \gamma_{zr} = \frac{\partial u_z}{\partial r} + \frac{\partial u_r}{\partial z} \\
&\varepsilon_z = \frac{\partial u_z}{\partial z}, \quad \gamma_{r\theta} = \frac{1}{r}\frac{\partial u_r}{\partial \theta} + \frac{\partial u_\theta}{\partial r} - \frac{u_\theta}{r}
\end{aligned}
\tag{1.1.5}
$$

The tensor form of the geometric equations in Cartesian coordinates can be written as

$$
\varepsilon_{ij} = \frac{1}{2}\left(u_{i,j} + u_{j,i}\right) \tag{1.1.6}
$$

with $2\varepsilon_{ij} = \gamma_{ij}$ when $i \neq j$.

The matrix form of the equations is given as

$$
\boldsymbol{\varepsilon} = \boldsymbol{E}^{\mathrm{T}}(\nabla)\boldsymbol{u} \tag{1.1.7}
$$

where $\boldsymbol{u} = [u_x, u_y, u_z]^{\mathrm{T}}$ and $\boldsymbol{E}(\nabla)$ is an operator matrix defined as

$$
\boldsymbol{E}(\nabla) = \begin{bmatrix}
\frac{\partial}{\partial x} & 0 & 0 & 0 & \frac{\partial}{\partial z} & \frac{\partial}{\partial y} \\
0 & \frac{\partial}{\partial y} & 0 & \frac{\partial}{\partial z} & 0 & \frac{\partial}{\partial x} \\
0 & 0 & \frac{\partial}{\partial z} & \frac{\partial}{\partial y} & \frac{\partial}{\partial x} & 0
\end{bmatrix}
\tag{1.1.8}
$$

1.1.2 Equations of Motion

When the governing equations of a dynamic problem are set at the equilibrium position, the body force is omitted in the equations. In Cartesian coordinates, we have the equations of motion as

$$
\begin{aligned}
&\frac{\partial \sigma_x}{\partial x} + \frac{\partial \tau_{yx}}{\partial y} + \frac{\partial \tau_{zx}}{\partial z} = \rho \frac{\partial^2 u_x}{\partial t^2} \\
&\frac{\partial \sigma_y}{\partial y} + \frac{\partial \tau_{xy}}{\partial x} + \frac{\partial \tau_{zy}}{\partial z} = \rho \frac{\partial^2 u_y}{\partial t^2} \\
&\frac{\partial \sigma_z}{\partial z} + \frac{\partial \tau_{xz}}{\partial x} + \frac{\partial \tau_{yz}}{\partial y} = \rho \frac{\partial^2 u_z}{\partial t^2}
\end{aligned}
\tag{1.1.9}
$$

where ρ is the density of the material.

Using Eq. (1.1.3), in cylindrical coordinates, the equations of motion become

$$\begin{aligned}
&\frac{\partial \sigma_r}{\partial r}+\frac{1}{r}\frac{\partial \tau_{r\theta}}{\partial \theta}+\frac{\partial \tau_{zr}}{\partial z}+\frac{\sigma_r-\sigma_\theta}{r}=\rho\frac{\partial^2 u_r}{\partial t^2}\\
&\frac{\partial \tau_{r\theta}}{\partial r}+\frac{1}{r}\frac{\partial \sigma_\theta}{\partial \theta}+\frac{\partial \tau_{\theta z}}{\partial z}+\frac{2\tau_{r\theta}}{r}=\rho\frac{\partial^2 u_\theta}{\partial t^2}\\
&\frac{\partial \tau_{zr}}{\partial r}+\frac{1}{r}\frac{\partial \tau_{\theta z}}{\partial \theta}+\frac{\partial \sigma_z}{\partial z}+\frac{\tau_{zr}}{r}=\rho\frac{\partial^2 u_z}{\partial t^2}
\end{aligned} \tag{1.1.10}$$

The tensor form of the equations of motion in Cartesian coordinates can be written as

$$\sigma_{ij,j}=\rho\ddot{u}_i \ (i=x,\ y,\ z \text{ or } r,\ \theta,\ z) \tag{1.1.11}$$

where the dots above the symbols denote the partial differentiation with respect to time t.

1.1.3 Constitutive Equations

The tensor form of Hooke's law in Cartesian coordinates can be written as

$$\sigma_{ij}=c_{ijkl}\,\varepsilon_{kl} \ (i,\ j,\ k,\ l=x,\ y,\ z \text{ or } r,\ \theta,\ z) \tag{1.1.12}$$

where c_{ijkl} are components of a fourth-rank tensor including 81 components. Since the stress vectors are symmetric, the exchange of the indices i and j does not alter the result. Noting that the strain vectors are symmetric as well, the same process can be done to the indices k and l; and then we have the relationships

$$c_{ijkl}=c_{jikl} \quad \text{and} \quad c_{ijkl}=c_{ijlk}$$

In addition, since we are considering the adiabatic process, we still have the following relationship:

$$c_{ijkl}=c_{klij}$$

Thus, among the 81 components of c_{ijkl}, the maximum number of independent ones is 21. For a homogeneous medium, the number of independent components goes down to 2, which are the Lame constants λ and μ, or Young's modulus E and Poisson's ratio v. And the tensor form of Hooke's law can be simplified as

$$\sigma_{ij}=2\mu\varepsilon_{ij}+\lambda\delta_{ij}\,\varepsilon_{kk} \tag{1.1.13}$$

where $\delta_{ij}=\begin{cases}1, & \text{when } i=j,\\ 0, & \text{when } i\neq j.\end{cases}$

Expanding Eq. (1.1.13), we have

$$\begin{aligned}\sigma_x &= \lambda e + 2\mu\varepsilon_x\\ \sigma_y &= \lambda e + 2\mu\varepsilon_y\\ \sigma_z &= \lambda e + 2\mu\varepsilon_z\\ \tau_{xy} &= \mu\gamma_{xy} = 2\mu\varepsilon_{xy}\\ \tau_{yz} &= \mu\gamma_{yz} = 2\mu\varepsilon_{yz}\\ \tau_{zx} &= \mu\gamma_{zx} = 2\mu\varepsilon_{zx}\end{aligned} \tag{1.1.14}$$

where λ and μ are Lame's constants with the relationship $\lambda=\dfrac{\nu E}{(1+\nu)(1-2\nu)}$. $e=\varepsilon_x+\varepsilon_y+\varepsilon_z$, which is called the volume strain or the matrix dilation.

The corresponding matrix form of the equations is given as

$$\boldsymbol{\sigma}=\boldsymbol{C}\boldsymbol{\varepsilon} \tag{1.1.15}$$

where $\boldsymbol{\sigma}$ and $\boldsymbol{\varepsilon}$ are vectors of stress and engineering strain, respectively. In Cartesian coordinates, they become

$$\boldsymbol{\sigma}=\left[\sigma_x,\ \sigma_y,\ \sigma_z,\ \gamma_{xy},\ \gamma_{yz},\ \gamma_{zx}\right]^{\mathrm{T}}$$

$$\boldsymbol{\varepsilon}=\left[\varepsilon_x,\ \varepsilon_y,\ \varepsilon_z,\ \gamma_{xy},\ \gamma_{yz},\ \gamma_{zx}\right]^{\mathrm{T}}$$

And $\boldsymbol{C}$ should be a nonsingular and reversible matrix, which can be written as

$$\boldsymbol{C}=\begin{bmatrix}\lambda+2\mu & \lambda & \lambda & 0 & 0 & 0\\ \lambda & \lambda+2\mu & \lambda & 0 & 0 & 0\\ \lambda & \lambda & \lambda+2\mu & 0 & 0 & 0\\ 0 & 0 & 0 & \mu & 0 & 0\\ 0 & 0 & 0 & 0 & \mu & 0\\ 0 & 0 & 0 & 0 & 0 & \mu\end{bmatrix} \tag{1.1.16}$$

1.2 Governing Equations of a Fully Saturated Poroelastic Medium

This section introduces the governing equations of a fully saturated poroelastic medium.

1.2.1 Governing Equations in Cartesian Coordinates

As for modeling the dynamic responses of a saturated porous medium, Biot[3] was the first to give the theory that presents three kinds of coupling (viscous, inertial, and mechanical) between the porous solid skeleton and pore fluid, and demonstrated that two kinds of longitudinal waves (P1 and P2 waves) and one kind of rational wave (the S wave) exist in the saturated porous medium. The existence of the P2 wave always distinguishes the two-phase medium from the single-phase one. The basic variables of Biot's theory are always the solid skeleton displacement (u) and the average displacement of pore fluid relative to the solid skeleton (w).

Considering the concept of effective stress of the saturated mixture, the relationship between effective stress, total stress and pore pressure can be expressed as

$$\sigma_{ij}'=\sigma_{ij}+\alpha\delta_{ij}p \tag{1.1.17}$$

where σ_{ij}' is the effective stress tensor, σ_{ij} is the total stress tensor, δ_{ij} is the Kronecker delta, and α is the Biot constant that depends on the geometry of material voids. For the most part, in soil mechanical problems, $\alpha \approx 1$ can be assumed. The relationship between total stress and effective stress becomes

$$\sigma_{ij}'=\sigma_{ij}+\delta_{ij}\,p \tag{1.1.18}$$

which corresponds to the classical effective stress definition by Terzaghi. Thus, the tensor form of the equations of motion for a fully saturated poroelastic medium becomes (omitting the body force)

$$\sigma_{ij,j}=\rho\ddot{u}_i+\rho_f\ddot{w}_i \tag{1.1.19}$$

where the dots above the symbols denote partial differentiation with respect to time t; thus, $\ddot{u}_i$ is the acceleration of the solid part, w_i is the fluid displacement relative to the solid part, and $\dot{w}_i$ is the fluid acceleration relative to the solid part. For fully saturated porous media (no air trapped inside), the density is equal to $\rho=n\rho_f+(1-n)\,\rho_s$, where n is the porosity, and ρ_s and ρ_f are the soil particle and water densities, respectively.

For the pore fluid, the equation of momentum balance can be expressed as

$$p_{,i}=-b\dot{w}_i-\rho_f\ddot{u}_i-m\ddot{w}_i \tag{1.1.20}$$

where w_i is the fluid velocity relative to the solid part, the parameter $b=\rho_f g/k_D$, where k_D is the Darcy permeability of the soil medium and g is the gravity; p is the pore water pressure.

According to the classical effective stress definition by Terzaghi in Eq. (1.1.17), the constitutive Eq. (1.1.13) becomes

$$\sigma_{ij}=2\mu\varepsilon_{ij}+\lambda\delta_{ij}\,\varepsilon_{kk}-\alpha\delta_{ij}p \tag{1.1.21}$$

The final equation is the mass conservation of the fluid flow, which is expressed by

$$\dot{p}=-\alpha M\dot{e}+M\dot{\varsigma} \tag{1.1.22}$$

where

$$\varsigma=-w_{i,i} \tag{1.1.23}$$

Substituting Eqs. (1.1.21)–(1.1.23) into Eqs. (1.1.19), (1.1.20), a ***u-w*** formulation can be obtained as follows:

$$\mu u_{i,jj}+(\lambda+\alpha^2M+\mu)\,u_{j,ji}+\alpha Mw_{j,ji}=\rho\ddot{u}_i+\rho_f\ddot{w}_i \tag{1.1.24}$$

$$\alpha Mu_{j,ji}+Mw_{j,ji}=\rho_f\ddot{u}_i+m\ddot{w}_i+\mathrm{b}\dot{w}_i \tag{1.1.25}$$

Zienkiewicz et al.[2] proposed a simplified ***u-p*** formulation in the context of finite element analysis of the liquefaction of saturated sand soil. By neglecting the second time derivatives of the relative fluid displacement from the original Biot[3] ***u-w*** formulation, the ***u-p*** formulation is deduced for reducing the primary variables in the sense of finite element calculation; there are 3 + 3 and 3 + 1 nodal variables in three-dimensional analysis for the ***u-w*** and ***u-p*** formulation, respectively. Moreover, the solid displacement (***u***) and the pore fluid pressure (*p*) are always the main concern.

By simply neglecting the pore fluid relative acceleration terms, i.e., $\ddot{w}_i$ in Eqs. (1.1.19), (1.1.20), it is easy to eliminate $\dot{w}_i$ using Eqs. (1.1.20), (1.1.22) and leaving u_i and p as primary variables. This simplified formulation is economical and convenient in numerical analysis, for there are only 3+1 nodal variables in the three-dimensional analysis. The equation set becomes

$$\sigma_{ij,j}=\rho\ddot{u}_i \tag{1.1.26}$$

$$\dot{u}_{i,i}+\left[-\frac{\rho_f}{b}\ddot{u}_i-\frac{1}{b}p_{,i}\right]_{,i}=\dot{p}/M \tag{1.1.27}$$

1.2.2 Governing Equations in Cylindrical Coordinates

In cylindrical coordinates, the expanded form of Eqs. (1.1.24), (1.1.25) become

$$
\begin{aligned}
&\mu\left(\nabla^2 u_r-\frac{1}{r^2}u_r-\frac{2}{r^2}\frac{\partial u_\theta}{\partial r}\right)+\left(\lambda+\alpha^2M+\mu\right)\frac{\partial e}{\partial r}-\alpha M\frac{\partial \varsigma}{\partial r}=\rho\ddot{u}_r+\rho_f\ddot{w}_r\\
&\mu\left(\nabla^2 u_\theta-\frac{1}{r^2}u_\theta-\frac{2}{r^2}\frac{\partial u_r}{\partial \theta}\right)+\left(\lambda+\alpha^2M+\mu\right)\frac{\partial e}{r\partial \theta}-\alpha M\frac{\partial \varsigma}{r\partial \theta}=\rho\ddot{u}_\theta+\rho_f\ddot{w}_\theta\\
&\mu\nabla^2 u_z+\left(\lambda+\alpha^2M+\mu\right)\frac{\partial e}{\partial z}-\alpha M\frac{\partial \varsigma}{\partial z}=\rho\ddot{u}_z+\rho_f\ddot{w}_z\\
&\alpha M\frac{\partial e}{\partial r}-M\frac{\partial \varsigma}{\partial r}=\rho_f\ddot{u}_r+m\ddot{w}_r+b\dot{w}_r\\
&\alpha M\frac{\partial e}{r\partial \theta}-M\frac{\partial \varsigma}{r\partial \theta}=\rho_f\ddot{u}_\theta+m\ddot{w}_\theta+b\dot{w}_\theta\\
&\alpha M\frac{\partial e}{\partial z}-M\frac{\partial \varsigma}{\partial z}=\rho_f\ddot{u}_z+m\ddot{w}_z+b\dot{w}_z
\end{aligned}
\tag{1.1.28}
$$

where e and ς are the matrix dilation and the fluid dilation relative to the solid, respectively, which are expressed in cylindrical coordinates as

$$
e=\frac{\partial u_r}{\partial r}+\frac{u_r}{r}+\frac{\partial u_\theta}{r\partial \theta}+\frac{\partial u_z}{\partial z},\quad \varsigma=-\left(\frac{\partial w_r}{\partial r}+\frac{w_r}{r}+\frac{\partial w_\theta}{r\partial \theta}+\frac{\partial w_z}{\partial z}\right)
$$

and ∇^2 denotes the Laplacian operator which is given by

$$
\nabla^2=\frac{\partial^2}{\partial r^2}+\frac{1}{r}\frac{\partial}{\partial r}+\frac{1}{r^2}\frac{\partial^2}{\partial \theta^2}+\frac{\partial^2}{\partial z^2}
$$

1.3 Boundary Conditions

In order to obtain a meaningful solution to a specific engineering problem, we need the external stimuli, called boundary conditions, to solve the governing equations of the fully saturated poroelastic medium.

The problem is in the area Ω with boundary Γ. For the total momentum balance on the part of the boundary Γ_N^t, we specify the total traction $\bar{T}_i$ (or in terms of the total stress, $\sigma_{ij}n_j$ with n_j being the jth component of the normal at the boundary) while for Γ_D^s, the displacement $\bar{u}_i$ is given.

For the fluid phase, again the boundary is divided into two parts, Γ_N^p on which the values of p are specified and Γ_D^w where the normal outflow w_i is prescribed (for instance, a zero value for the normal outward velocity on an impermeable boundary).

Summarizing, for the overall assembly, we can thus write

Compulsive boundary conditions

$$
\begin{cases}
u_i=\bar{u}_i \text{ on } \Gamma=\Gamma_D^s\\
w_i=\bar{w}_i \text{ on } \Gamma=\Gamma_D^w
\end{cases}
\tag{1.1.29}
$$

Natural boundary conditions

$$\begin{cases} \sigma_{ij}n_j = \bar{T}_i \text{ on } \Gamma = \Gamma_N^t \\ pn_i = \bar{p}_n \text{ on } \Gamma = \Gamma_N^p \end{cases} \tag{1.1.30}$$

The relationships between the two boundaries

$$\begin{cases} \Gamma = \Gamma_D^s \cup \Gamma_D^w \cup \Gamma_N^t \cup \Gamma_N^p \\ \Gamma_D^s \cap \Gamma_N^t = \varnothing \\ \Gamma_D^w \cap \Gamma_N^p = \varnothing \end{cases} \tag{1.1.31}$$

Chapter 2
Solutions for Saturated Soil Under Moving Loads and Engineering Applications

This chapter introduces solutions for saturated ground governed by Biot's theory to a moving rectangular pressure. Furthermore, the applications of Biot's theory to the investigation of dynamic responses of highway roads and railway tracks subjected to a moving traffic load are presented in Sections 2.1 and 2.2, respectively.

2.1 General Solutions for Saturated Ground Under a Moving Load

The governing equations are given in Eqs. (1.1.24), (1.1.25). The constitutive relationships are given in Eqs. (1.1.21), (1.1.22).

2.1.1 *Solutions for Saturated Ground Generated by a Moving Rectangular Pressure*

Fourier transforms with respect to time t are defined as

$$\tilde{f}(x, y, z, \omega) = \int_{-\infty}^{+\infty} f(x, y, z, t)\mathrm{e}^{-\mathrm{i}\omega t}\mathrm{d}t \tag{2.1.1}$$

$$f(x, y, z, t) = \frac{1}{2\pi}\int_{-\infty}^{\infty} \tilde{f}(x, y, z, \omega)\mathrm{e}^{\mathrm{i}\omega t}\mathrm{d}\omega \tag{2.1.2}$$

Fourier transform pairs with respect to x and y are defined as

$$\overline{\overline{\tilde{f}}}(\xi, \eta, z, a_0) = \int_{-\infty}^{\infty}\int_{-\infty}^{\infty} \tilde{f}(x, y, z, a_0)\mathrm{e}^{-\mathrm{i}(\xi x + \eta y)}\mathrm{d}x\mathrm{d}y \tag{2.1.3}$$

$$\tilde{f}(x, y, z, a_0) = \frac{1}{4\pi^2}\int_{-\infty}^{\infty}\int_{-\infty}^{\infty} \overline{\overline{\tilde{f}}}(\xi, \eta, z, a_0)\mathrm{e}^{\mathrm{i}(\xi x + \eta y)}\mathrm{d}\xi\mathrm{d}\eta \tag{2.1.4}$$

By using the Fourier transform given in Eq. (2.1.1), (1.1.24), (1.1.25), (1.1.21), and (1.1.22) are transformed from partial differential equations to ordinary differential equations. The governing equations in the x direction are given as

$$\mu\nabla^2\tilde{u}_x + (\lambda+\alpha^2M+\mu)\frac{\partial\tilde{\theta}}{\partial x} - \alpha M\frac{\partial\tilde{\zeta}}{\partial x} = -\omega^2\rho\tilde{u}_x - \omega^2\rho_f\tilde{w}_x \tag{2.1.5}$$

$$\alpha M\frac{\partial\tilde{\theta}}{\partial x} - M\frac{\partial\tilde{\zeta}}{\partial x} = -\omega^2\rho_f\tilde{u}_x - \omega^2 m\tilde{w}_x + \mathrm{i}\omega b\tilde{w}_x \tag{2.1.6}$$

$$\frac{\partial\tilde{p}}{\partial x} = M\frac{\partial\tilde{\zeta}}{\partial x} - \alpha M\frac{\partial\tilde{\theta}}{\partial x} \tag{2.1.7}$$

Substituting Eq. (2.1.7) into Eq. (2.1.5) yields

$$\mu\nabla^2\tilde{u}_x + (\lambda+\mu)\frac{\partial\tilde{\theta}}{\partial x} - \alpha\frac{\partial\tilde{p}_f}{\partial x} = -\omega^2\rho\tilde{u}_x - \omega^2\rho_f\tilde{w}_x \tag{2.1.8}$$

Substituting Eq. (2.1.7) into Eq. (2.1.6) gives

$$-\frac{\partial\tilde{p}_f}{\partial x} = -\omega^2\rho_f\tilde{u}_x - \omega^2 m\tilde{w}_x + \mathrm{i}\omega b\tilde{w}_x \tag{2.1.9}$$

and then the following equation can be obtained from Eq. (2.1.9):

$$\tilde{w}_x = \frac{\omega^2\rho_f\tilde{u}_x - \dfrac{\partial\tilde{p}_f}{\partial x}}{\mathrm{i}\omega b - \omega^2 m} \tag{2.1.10}$$

Then substituting Eq. (2.1.10) into Eq. (2.1.9) yields

$$\mu\nabla^2\tilde{u}_x + (\lambda+\mu)\frac{\partial\tilde{\theta}}{\partial x} + \omega^2(\rho-\vartheta\rho_f)\tilde{u}_x - (\alpha-\vartheta)\frac{\partial\tilde{p}_f}{\partial x} = 0 \tag{2.1.11}$$

where $\vartheta = \dfrac{\omega^2\rho_f}{\omega^2 m - \mathrm{i}\omega b}$.

Similarly, the following equations in the y and z directions can be obtained:

$$\mu\nabla^2\tilde{u}_y+(\lambda+\mu)\frac{\partial\tilde{\theta}}{\partial y}+\omega^2(\rho-\vartheta\rho_f)\tilde{u}_y-(\alpha-\vartheta)\frac{\partial\tilde{p}_f}{\partial y}=0 \tag{2.1.12}$$

$$-\frac{\partial\tilde{p}_f}{\partial y}=-\omega^2\rho_f\tilde{u}_y-\omega^2m\tilde{w}_y+\mathrm{i}\omega b\tilde{w}_y \tag{2.1.13}$$

$$\mu\nabla^2\tilde{u}_z+(\lambda+\mu)\frac{\partial\tilde{\theta}}{\partial z}+\omega^2(\rho-\vartheta\rho_f)\tilde{u}_z-(\alpha-\vartheta)\frac{\partial\tilde{p}_f}{\partial z}=0 \tag{2.1.14}$$

$$-\frac{\partial\tilde{p}_f}{\partial z}=-\omega^2\rho\tilde{u}_z-\omega^2m\tilde{w}_z+\mathrm{i}\omega b\tilde{w}_z \tag{2.1.15}$$

Taking the derivative of Eqs. (2.1.9), (2.1.13), (2.1.15) with respect to *x*, *y*, *z*, respectively, and adding them together yields

$$-\nabla^2\tilde{p}_f=-\omega^2\rho_f\tilde{\theta}+\omega^2m\tilde{\zeta}-\mathrm{i}\omega b\tilde{\zeta} \tag{2.1.16}$$

Eq. (1.1.22) in the transformed domain can be expressed as

$$\tilde{p}_f=M\tilde{\zeta}-\alpha M\tilde{\theta} \tag{2.1.17}$$

Substituting Eq. (2.1.17) into Eq. (2.1.16) yields

$$\nabla^2\tilde{p}_f+\frac{\omega^2\rho_f}{\vartheta M}\tilde{p}_f+\frac{\omega^2\rho_f(\alpha-\vartheta)}{\vartheta}\tilde{\theta}=0 \tag{2.1.18}$$

By applying the operation ∇^2 to Eq. (2.1.18), the following equation can be obtained:

$$\nabla^4\tilde{p}_f+\frac{a_0^2\rho_f}{\vartheta M}\nabla^2\tilde{p}_f+\frac{a_0^2\rho_f(\alpha-\vartheta)}{\vartheta}\nabla^2\tilde{\theta}=0 \tag{2.1.19}$$

With the aid of Eq. (2.1.18), the following equation can then be obtained:

$$\tilde{\theta}=\frac{M\nabla^2\tilde{p}_f+(\omega^2m-\mathrm{i}\omega b)\tilde{p}_f}{\omega^2\rho_fM-\alpha M(\omega^2m-\mathrm{i}\omega^2b)} \tag{2.1.20}$$

Taking the derivative of Eqs. (2.1.11), (2.1.12), (2.1.14) with respect to *x*, *y*, *z*, respectively, and adding them together, the following equations can be obtained:

$$(\lambda+2\mu)\nabla^2\tilde{\theta}+\omega^2(\rho-\vartheta\rho_f)\tilde{\theta}-(\alpha-\vartheta)\nabla^2\tilde{p}_f=0 \tag{2.1.21}$$

and thus

$$\nabla^2\tilde{\theta} = \frac{(\alpha-\vartheta)\nabla^2\tilde{p}_f - \omega^2(\rho-\vartheta\rho_f)\tilde{\theta}}{\lambda+2\mu} \tag{2.1.22}$$

By substituting Eqs. (2.1.20), (2.1.22) into Eq. (2.1.19), the following equation can be obtained:

$$\nabla^4\tilde{p}_f + \beta_1\nabla^2\tilde{p}_f + \beta_2\tilde{p}_f = 0 \tag{2.1.23}$$

where

$$\beta_1 = \frac{(\omega^2 m - \mathrm{i}\omega b)(\lambda+\alpha^2 M+2\mu) + \omega^2 M\rho - 2\alpha\omega^2 M\rho_f}{(\lambda+2\mu)M} \tag{2.1.24}$$

$$\beta_2 = \frac{\rho\omega^2(\omega^2 m - \mathrm{i}\omega b) - \omega^4\rho_f^2}{(\lambda+2\mu)M} \tag{2.1.25}$$

Applying Fourier transforms with respect to x and y, Eq. (2.1.23) is transformed from a partial differential equation to an ordinary differential equation:

$$\frac{\mathrm{d}^4\bar{\bar{\tilde{p}}}_f}{\mathrm{d}z^4} + \left(\beta_1 - 2\xi^2 - 2\eta^2\right)\frac{\mathrm{d}^2\bar{\bar{\tilde{p}}}_f}{\mathrm{d}z^2} + \left(\xi^4+\eta^4+2\xi^2\eta^2 - \beta_1\xi^2 - \beta_1\eta^2 + \beta_2\right)\bar{\bar{\tilde{p}}}_f = 0 \tag{2.1.26}$$

By solving Eq. (2.1.26), the expression $\bar{\bar{\tilde{p}}}_f$ can be obtained:

$$\bar{\bar{\tilde{p}}}_f = A\mathrm{e}^{-\gamma_1 z} + B\mathrm{e}^{-\gamma_2 z} \tag{2.1.27}$$

At $z \to \infty$, the dynamic responses become zero; therefore $\mathrm{Re}(\gamma_i) > 0$, $i=1,2$, where

$$\gamma_1 = \sqrt{\xi^2+\eta^2-L_1^2},\quad \gamma_2 = \sqrt{\xi^2+\eta^2-L_2^2} \tag{2.1.28}$$

$$L_1^2 = \frac{\beta_1 + \sqrt{\beta_1^2 - 4\beta_2}}{2},\quad L_2^2 = \frac{\beta_1 - \sqrt{\beta_1^2 - 4\beta_2}}{2} \tag{2.1.29}$$

By using the Fourier transforms given in Eq. (2.1.3) in Eq. (2.1.17), the following equation can be obtained:

$$\left(-\xi^2 - \eta^2 + \frac{\mathrm{d}^2}{\mathrm{d}z^2}\right)\bar{\bar{\tilde{p}}}_f + \frac{\omega^2\rho_f}{\vartheta M}\bar{\bar{\tilde{p}}}_f + \frac{\omega^2\rho_f(\alpha-\vartheta)}{\vartheta}\bar{\bar{\tilde{\theta}}} = 0 \tag{2.1.30}$$

Substituting Eq. (2.1.27) into Eq. (2.1.30) yields

$$-\frac{\omega^2\rho_{\mathrm{f}}(\alpha-\vartheta)}{\vartheta}\bar{\bar{\theta}}=\left(\gamma_1^2-\xi^2-\eta^2+\frac{\omega^2\rho_{\mathrm{f}}}{\vartheta M}\right)A\mathrm{e}^{-\gamma_1 z}+\left(\gamma_2^2-\xi^2-\eta^2+\frac{\omega^2\rho_{\mathrm{f}}}{\vartheta M}\right)B\mathrm{e}^{-\gamma_2 z} \tag{2.1.31}$$

and then

$$\bar{\bar{\theta}}=A\chi_1\mathrm{e}^{-\gamma_1 z}+B\chi_2\mathrm{e}^{-\gamma_2 z} \tag{2.1.32}$$

where

$$\chi_1=\frac{\vartheta ML_1^2-\omega^2\rho_{\mathrm{f}}}{\omega^2\rho_{\mathrm{f}}M(\alpha-\vartheta)},\quad \chi_2=\frac{\vartheta ML_2^2-\omega^2\rho_{\mathrm{f}}}{\omega^2\rho_{\mathrm{f}}M(\alpha-\vartheta)} \tag{2.1.33}$$

Applying Fourier transforms to Eq. (2.1.14) with respect to x and y and then obtain

$$\mu\left(-\xi^2-\eta^2+\frac{\mathrm{d}^2}{\mathrm{d}z^2}\right)\bar{\bar{u}}_z+(\lambda+\mu)\frac{\partial\bar{\bar{\theta}}}{\partial z}+a_0^2(\rho-\vartheta\rho_{\mathrm{f}})\bar{\bar{u}}_z-(\alpha-\vartheta)\frac{\partial\bar{\bar{p}}_{\mathrm{f}}}{\partial z}=0 \tag{2.1.34}$$

By substituting Eqs. (2.1.27), (2.1.32) into Eq. (2.1.34),

$$\begin{aligned}\mu\frac{\mathrm{d}^2\bar{\bar{u}}_z}{\mathrm{d}z^2}&+\left(S^2-\mu\xi^2-\mu\eta^2\right)\bar{\bar{u}}_z-(\lambda+\mu)\left(\gamma_1A\chi_1\mathrm{e}^{-\gamma_1 z}+\gamma_2B\chi_2\mathrm{e}^{-\gamma_2 z}\right)\\&+(\alpha-\vartheta)\left(\gamma_1A\mathrm{e}^{-\gamma_1 z}+\gamma_2B\mathrm{e}^{-\gamma_2 z}\right)=0\end{aligned} \tag{2.1.35}$$

where

$$S^2=\omega^2\,(\rho-\vartheta\rho_{\mathrm{f}}) \tag{2.1.36}$$

The general solution for Eq. (2.1.35) is $\bar{\bar{u}}_z=Aa_1\gamma_1\mathrm{e}^{-\gamma_1 z}+Ba_2\gamma_2\mathrm{e}^{-\gamma_2 z}$ and the particular solution is $\bar{\bar{u}}_z=C\mathrm{e}^{-\gamma_3 z}$; then the solution of $\bar{\bar{u}}_z$ can be expressed as

$$\bar{\bar{u}}_z=Aa_1\gamma_1\mathrm{e}^{-\gamma_1 z}+Ba_2\gamma_2\mathrm{e}^{-\gamma_2 z}+C\mathrm{e}^{-\gamma_3 z} \tag{2.1.37}$$

where

$$\gamma_3=\sqrt{\xi^2+\eta^2-S^2/\mu},\quad \mathrm{Re}(\gamma_3)>0 \tag{2.1.38}$$

$$a_1=\frac{\chi_1(\lambda+\mu)-\alpha+\vartheta}{S^2-\mu L_1^2},\quad a_2=\frac{\chi_2(\lambda+\mu)-\alpha+\vartheta}{S^2-\mu L_2^2} \tag{2.1.39}$$

By applying Fourier transforms with respect to x and y, Eq. (2.1.12) can be expressed as

$$\mu\frac{\mathrm{d}^2\bar{\bar{u}}_y}{\mathrm{d}z^2}+\left(S^2-\mu\xi^2-\mu\eta^2\right)\bar{\bar{u}}_y+\mathrm{i}\eta[(\lambda+\mu)(A\chi_1\mathrm{e}^{-\gamma_1 z}+B\chi_2\mathrm{e}^{-\gamma_2 z}) -(\alpha-\vartheta)(A\mathrm{e}^{-\gamma_1 z}+B\mathrm{e}^{-\gamma_2 z})]=0 \tag{2.1.40}$$

and thus

$$\bar{\bar{u}}_y=-\mathrm{i}\eta(Aa_1\mathrm{e}^{-\gamma_1 z}+Ba_2\mathrm{e}^{-\gamma_2 z})+\mathrm{i}D\mathrm{e}^{-\gamma_3 z} \tag{2.1.41}$$

Applying Fourier transforms with respect to x and y into $\theta=u_{i,i}$, the following equation can be obtained:

$$\bar{\bar{\theta}}=\mathrm{i}\xi\bar{\bar{u}}_x+\mathrm{i}\eta\bar{\bar{u}}_y+\frac{\partial\bar{\bar{u}}_z}{\partial z} \tag{2.1.42}$$

By substituting Eqs. (2.1.37), (2.1.41) into Eq. (2.1.42), the solutions for $\bar{\bar{u}}_z$ can be obtained as

$$\bar{\bar{u}}_x=-\frac{\mathrm{i}}{\xi}\left\{\left[\chi_1+a_1\left(\xi^2-L_1^2\right)\right]A\mathrm{e}^{-\gamma_1 z}+\left[\chi_2+a_2\left(\xi^2-L_2^2\right)\right]B\mathrm{e}^{-\gamma_2 z}+(C\gamma_3+D\eta)\mathrm{e}^{-\gamma_3 z}\right\} \tag{2.1.43}$$

The constitutive equation in saturated soil can be expressed as

$$\sigma_z=\lambda\theta+2\mu\frac{\partial u_z}{\partial z}-\alpha p_{\mathrm{f}} \tag{2.1.44}$$

$$\tau_{xz}=\mu\left(\frac{\partial u_x}{\partial z}+\frac{\partial u_z}{\partial x}\right) \tag{2.1.45}$$

$$\tau_{yz}=\mu\left(\frac{\partial u_y}{\partial z}+\frac{\partial u_z}{\partial y}\right) \tag{2.1.46}$$

The expressions in the transformed domain can be expressed as

$$\bar{\bar{\sigma}}_z=\lambda\bar{\bar{\theta}}+2\mu\frac{\partial\bar{\bar{u}}_z}{\partial z}-\alpha\bar{\bar{p}}_{\mathrm{f}} \tag{2.1.47}$$

$$\bar{\bar{\tau}}_{xz}=\mu\frac{\partial\bar{\bar{u}}_x}{\partial z}+\mathrm{i}\xi\mu\bar{\bar{u}}_z \tag{2.1.48}$$

$$\bar{\bar{\tau}}_{yz}=\mu\frac{\partial\bar{\bar{u}}_y}{\partial z}+\mathrm{i}\eta\mu\bar{\bar{u}}_z \tag{2.1.49}$$

Substituting Eqs. (2.1.27), (2.1.32), (2.1.37), (2.1.41), (2.1.43) into Eqs. (2.1.47)–(2.1.49) yields

$$\overline{\overline{\overline{\sigma}}}_z = g_3 A e^{-\gamma_1 z} + g_4 B e^{-\gamma_2 z} - 2\mu\gamma_3 C e^{-\gamma_3 z} \tag{2.1.50}$$

$$\overline{\overline{\overline{\tau}}}_{xz} = \frac{\mu i}{\xi}\left[g_1\gamma_1 A e^{-\gamma_1 z} + g_2\gamma_2 B e^{-\gamma_2 z} + \left(C\gamma_3^2 + C\xi^2 + \eta\gamma_3 D\right)e^{-\gamma_3 z}\right] \tag{2.1.51}$$

$$\overline{\overline{\overline{\tau}}}_{yz} = \mu i[2\eta a_1\gamma_1 A e^{-\gamma_1 z} + 2\eta a_2\gamma_2 B e^{-\gamma_2 z} + (C\eta - D\gamma_3)e^{-\gamma_3 z}] \tag{2.1.52}$$

where

$$g_1=\chi_1+a_1\ (2\xi^2-L_1^{\ 2}) \tag{2.1.53}$$

$$g_2=\chi_2+a_2\ (2\xi^2-L_2^{\ 2}) \tag{2.1.54}$$

$$g_3=\chi_1\lambda-2\mu a_1\ \gamma_1^{\ 2}-\alpha \tag{2.1.55}$$

$$g_4=\chi_2\lambda-2\mu a_2\ \gamma_2^{\ 2}-\alpha \tag{2.1.56}$$

A uniformly distributed rectangular pressure is applied to the surface of the poroelastic half-space. Thus, the boundary conditions on the surface of the half-space (z = 0 m) are given as τ_{xz}=0 kPa, τ_{yz}=0 kPa, p_f=0 kPa , and the vertical stress is expressed as

$$\sigma_z(x, y, t) = \begin{cases} P, |x - Vt| \le a, |y| \le b \\ 0, |x - Vt| > a, |y| > b \end{cases} \tag{2.1.57}$$

The vertical stress in the transformed domain can be expressed as

$$\overline{\overline{\overline{\sigma}}}_z(\xi, \eta, a_0) = \begin{cases} \dfrac{8\pi P}{\xi\eta}\sin\xi a \cdot \sin\eta b \cdot \delta(\xi V + a_0),\ \xi\eta \ne 0 \\ \dfrac{8\pi P}{\eta} a \cdot \sin\eta b \cdot \delta(a_0),\ \xi = 0, \eta \ne 0 \\ \dfrac{8\pi P}{\xi} b \sin\xi a \cdot \delta(\xi V + a_0),\ \xi \ne 0, \eta = 0 \\ 8\pi Pab \cdot \delta(a_0),\ \xi = 0, \eta = 0 \end{cases} \tag{2.1.58}$$

By substituting the boundary condition into Eqs. (2.1.27), (2.1.50)–(2.1.52), and introducing the auxiliary spatial coordinate x_t=x–Vt, the solution in the time domain is obtained:

$$\begin{aligned} u_z(x_t, y, z) &= \frac{1}{(2\pi)^3}\int_{-\infty}^{+\infty}\int_{-\infty}^{+\infty}\int_{-\infty}^{+\infty}\overline{\overline{\overline{\sigma}}}_z(\xi, \eta, \omega)\phi(\xi, \eta, z, \omega)e^{i\xi x}e^{i\eta y}e^{ia_0 t}\,d\xi\, d\eta\, d\omega \\ &= \frac{1}{(2\pi)^2}\int_{-\infty}^{+\infty}\int_{-\infty}^{+\infty}\overline{\overline{\overline{\sigma}}}_z(\xi, \eta)\phi(\xi, \eta, z, -\xi V)e^{i\xi x_t}e^{i\eta y}\,d\xi\, d\eta \end{aligned} \tag{2.1.59}$$

where

$$\phi(\xi,\eta,z,\alpha_0)=\frac{1}{\Delta_0}\left[\left(\gamma_3^2+\xi^2+\eta^2\right)\left(\gamma_2 a_2 e^{-\gamma_2 z}-\gamma_1 a_1 e^{-\gamma_1 z}\right)+\left(\gamma_1 g_5-\gamma_2 g_6\right)e^{-\gamma_3 z}\right] \tag{2.1.60}$$

$$\Delta_0=(g_4-g_3)\left(\gamma_3^2+\xi^2+\eta^2\right)-2\mu\gamma_3(\gamma_1 g_5-\gamma_2 g_6) \tag{2.1.61}$$

2.2 Application in Highway Engineering

With increasing traffic volume and traffic running speeds, the serviceability of highway roads is increasingly influenced by the traffic. Road traffic loads cause distresses such as cracks and settlement. Furthermore, due to the low Rayleigh wave velocity in soft soils, vehicles can possibly approach the Rayleigh wave velocity in the poroelastic half-space. Vibrations and stresses in the pavement structure and the ground will be amplified in this circumstance, causing disturbances to residences along the line. It is of great significance to predict the dynamic responses of pavement-ground systems to the moving load using an accurate model.

2.2.1 Governing Equations for the Pavement Structure

The dynamic response of a pavement on a saturated poroelastic half-space was first investigated by Cai et al.[4] The pavement structure was simplified as a thin plate extending to infinity in the horizontal direction. The model of pavement on the saturated ground is shown in Fig. 2.2.1.

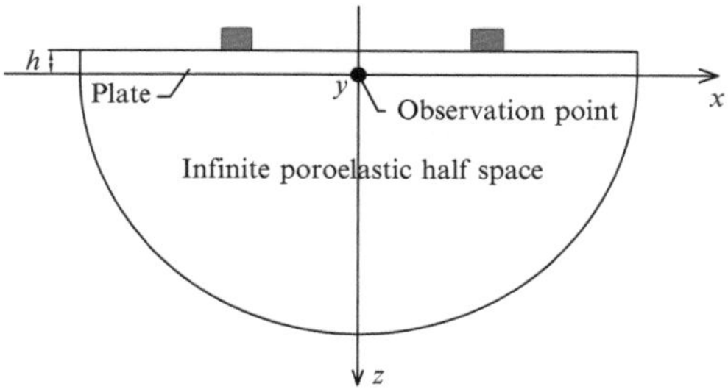

Fig. 2.2.1 Theoretical model of the plate-ground system

The Kirchhoff small-deflection thin-plate theory was employed as by Kim and Roessert[5]. Denoting the reaction of the ground to the plate by $F(x, y, t)$, the governing equation of the plate is given as follows:

$$D_p\left[\frac{\partial^4 w(x,y,t)}{\partial x^4}+2\frac{\partial^4 w(x,y,t)}{\partial x^2\partial y^2}+\frac{\partial^4 w(x,y,t)}{\partial y^4}\right]+m_b\frac{\partial^2 w(x,y,t)}{\partial t^2}=q(x,y,t)+F(x,y,t) \tag{2.2.1}$$

where w is the vertical displacement of the plate, m_b is the mass density of the plate per unit area, and $q(x, y, t)$ is the traffic load pressure acting on the surface of the road. D_p is the flexural rigidity of the plate and can be defined as follows:

$$D_p = \frac{Eh^3}{12(1-\nu_p{}^2)} \tag{2.2.2}$$

where E, h, and ν_p are the elastic modulus, thickness, and Poisson's ratio of the plate, respectively.

The traffic load is modeled as four rectangular load pressures expressed as

$$q(x, y, t) = \begin{cases} Q, \dfrac{d_a - d_1}{2} \leq |x - Vt| \leq \dfrac{d_a + d_1}{2} \text{ and } \dfrac{d_w - d_2}{2} \leq |y| \leq \dfrac{d_w + d_2}{2} \\ 0, \text{ otherwise} \end{cases} \tag{2.2.3}$$

where Q is the magnitude of the load pressure; d_1 and d_2 are the loaded lengths of a tire print in the x and y directions, respectively; d_w is the distance between the left and right wheels; and d_a is the distance between the front and rear wheels. V is the vehicle speed.

The constitutive equations of the thin plate are given as

$$\sigma_x = \frac{E}{1-\nu_p^2}\left(\frac{\partial^2 w}{\partial x^2} + \nu_p \frac{\partial^2 w}{\partial y^2}\right) \tag{2.2.4}$$

$$\sigma_y = \frac{E}{1-\nu_p^2}\left(\frac{\partial^2 w}{\partial y^2} + \nu_p \frac{\partial^2 w}{\partial x^2}\right) \tag{2.2.5}$$

2.2.2 Coupling Between the Pavement and Saturated Ground

As the contact surface between the pavement and the poroelastic half-space is assumed to be smooth and fully permeable, the boundary conditions of the half-space at z=0 m are given as follows:

$$\sigma_z (x, y, 0, t) = -F(x, y, t) \tag{2.2.6}$$

$$\tau_{xz} (x, y, 0, t) = 0 \tag{2.2.7}$$

$$\tau_{yz} (x, y, 0, t) = 0 \tag{2.2.8}$$

$$p_f (x, y, 0, t) = 0 \tag{2.2.9}$$

$$u_z (x, y, 0, t) = w(x, y, t) \tag{2.2.10}$$

2.2.3 Solution Methods for the Pavement and Saturated Ground

The solutions of the plate-ground system were obtained based on the Fourier transform. Applying a triple Fourier transform with respect to x, y, and t, Eq. (2.2.1) yields

$$\left[D_{\mathrm{p}}\left(\varepsilon^2+\eta^2\right)^2-m_{\mathrm{b}}\omega^2\right]\overline{\overline{\overline{w}}}=\overline{\overline{\overline{q}}}(\varepsilon,\eta,\omega)+\overline{\overline{\overline{F}}}(\varepsilon,\eta,\omega) \tag{2.2.11}$$

The traffic load in the transformed domain can be expressed as

$$\overline{\overline{\overline{q}}}(\varepsilon,\eta,\omega)=\begin{cases}8\pi Q\dfrac{\sin\dfrac{d_1\varepsilon}{2}\sin\dfrac{d_2\eta}{2}}{\varepsilon\eta}\mathrm{e}^{-\frac{d_{\mathrm{a}}}{2}\varepsilon\mathrm{i}}\mathrm{e}^{-\frac{d_{\mathrm{w}}}{2}\eta\mathrm{i}}\left(1+\mathrm{e}^{d_a\varepsilon\mathrm{i}}\right)\left(1+\mathrm{e}^{d_w\eta\mathrm{i}}\right)\delta(\omega+\varepsilon V), & \varepsilon\eta\neq0\\ 8\pi Qd_1d_2\delta(\omega), & \varepsilon=0,\eta=0\end{cases} \tag{2.2.12}$$

Combined with the boundary conditions presented by Eqs. (2.2.6)–(2.2.10) and the solutions of the saturated ground presented in Section 2.1, the solutions of the plate-ground system can be derived as

$$\begin{aligned}&u_z(x,y,z,t)\\&=\frac{1}{(2\pi)^3}\int_{-\infty}^{\infty}\int_{-\infty}^{\infty}\int_{-\infty}^{\infty}\overline{\overline{\overline{q}}}(\varepsilon,\eta,\omega)\frac{\phi(\varepsilon,\eta,z,\omega)}{\left[D_{\mathrm{p}}(\varepsilon^2+\eta^2)^2-m_{\mathrm{b}}\omega^2\right]\phi(\varepsilon,\eta,0,\omega)+1}\mathrm{e}^{\mathrm{i}\varepsilon x}\mathrm{e}^{\mathrm{i}\eta y}\mathrm{e}^{\mathrm{i}\omega\tau}\mathrm{d}\varepsilon\mathrm{d}\eta\mathrm{d}\omega\end{aligned} \tag{2.2.13}$$

where

$$\phi(\varepsilon,\eta,z,\omega)=\frac{1}{\Delta_0}\left[\left({\gamma_3}^2+\varepsilon^2+\eta^2\right)\left(\gamma_2a_2\mathrm{e}^{-\gamma_2z}-\gamma_1a_1\mathrm{e}^{-\gamma_1z}\right)+\left(\gamma_1g_5-\gamma_2g_6\right)\mathrm{e}^{-\gamma_3z}\right] \tag{2.2.14}$$

$$g_5=g_1+2\eta^2a_1,\quad g_6=g_1+2\eta^2a_2 \tag{2.2.15}$$

$$\Delta_0=\left(g_4-g_3\right)\left(\gamma_3^2+\varepsilon^2+\eta^2\right)-2\gamma_3\left(\gamma_1g_5-\gamma_2g_6\right) \tag{2.2.16}$$

With the nontrivial properties of the delta function and introducing auxiliary spatial coordinate $x_t=x-Vt$, Eq. (2.2.13) can be written as

$$u_z(x_{\mathrm{t}},y,z)=\frac{1}{(2\pi)^2}\int_{-\infty}^{\infty}\int_{-\infty}^{\infty}\overline{\overline{q}}(\varepsilon,\eta)\frac{\phi(\varepsilon,\eta,z,-\varepsilon c_0)\mathrm{e}^{\mathrm{i}[\varepsilon x_{\mathrm{t}}+\eta y]}}{\left[D_{\mathrm{p}}(\varepsilon^2+\eta^2)^2-m_{\mathrm{b}}\varepsilon^2V^2\right]\phi(\varepsilon,\eta,0,-\varepsilon V)+1}\mathrm{d}\varepsilon\mathrm{d}\eta \tag{2.2.17}$$

2.2.4 Numerical Results and Conclusions

Dimensionless parameters were adopted in the numerical analysis to demonstrate the general trends in the numerical results. The parameters were nondimensionalized through $x=x_1/a$, $y=y_1/a$, $z=z_1/a$,

$\tau=t/a, \sqrt{\mu/\rho}$, $\lambda^*=\dfrac{\lambda}{\mu}$, $M^*=\dfrac{M}{\mu}$, $\rho^*=\dfrac{\rho_f}{\rho}$, $m^*=\dfrac{m}{\rho}$, $b^*=\dfrac{ab}{\sqrt{\rho\mu}}$, where a is the unit length. The dimensionless parameters for fully saturated ground are presented in Table 2.2.1. The dimensionless parameters of the plate are presented in Table 2.2.2. The dimensionless parameters of the vehicle can be found in Table 2.2.3.

Table 2.2.1 Dimensionless parameters for fully water-saturated poroelastic soil medium

Parameter	Value
Shear modulus of solid, μ^*	1
Poisson's ratio, v	0.35
Solid density, ρ_s^*	1.219
Water density, ρ_w^*	0.671
Porosity, n	0.4
Coefficient of material damping, D	0.05
Ratio between the fluid viscosity and the intrinsic permeability, b^*	0.1–100
Bulk modulus of the fluid, K_f^*	1.299

Table 2.2.2 Dimensionless parameters of the plate

Parameter	Value
Elastic modulus, E^*	454.55
Poisson's ratio, v_p	0.15
Plate thickness, h^*	0.30
Mass density, m_b^*	0.47
Damping ratio, D_1	0

Table 2.2.3 Dimensionless parameters of tandem-axle loads

Parameter	Value
Load pressure, Q*	0.013
Load length, d_1*	0.2
Load width, d_2*	0.2
Distance between left and right wheels, d_w*	1.5
Distance between front and rear wheels, d_a*	3

The vertical displacements at the surface of the pavement between are studied in Fig. 2.2.2 considering four different b^* values. With the assumption of a constant fluid viscosity value, then the intrinsic permeability of the soil medium decreases as b increases. In Fig. 2.2.2A for c_0=0.2, the displacement curves are almost symmetric and have one peak at the center of the tandem axles. The maximum displacement decreases with increasing b^*. In Fig. 2.2.2B for the high load velocity of c_0=1.0, the displacements become much larger than those of c_0=0.2, and significant ground vibration

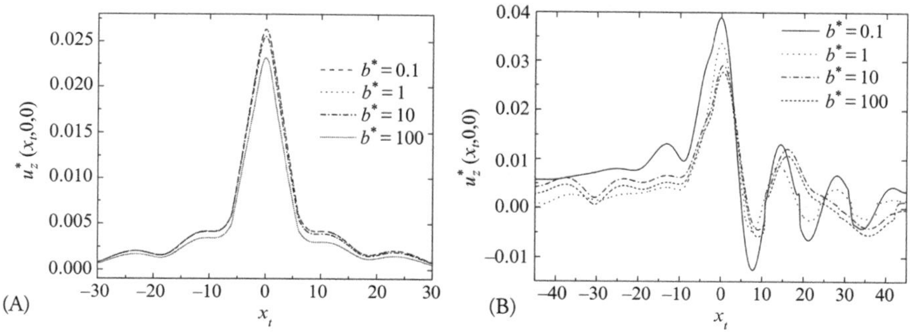

Fig. 2.2.2 Effects of the soil permeability coefficients on the displacement responses. (A) c_0=0.2; (B) c_0 =1.0

is observed. The maximum displacements apparently decrease as b^* increases.

The dimensionless maximum vertical displacement against dimensionless load velocity for different values of flexural rigidity is plotted in Fig. 2.2.3. The displacements are normalized with respect to the displacement at c_0=0. The maximum surface displacement response has nearly a static character up to a velocity of c_0=0.5. However, with a further increase in load velocity, the displacements increase rapidly and reach a maximum value near c_0=0.93 and then decrease rapidly. The critical speed for the poroelastic half-space is the Rayleigh wave speed, which equals 0.936V_s with a soil Poisson's ratio of 0.35 according to Richart[6]. Thus, the critical speed of the pavement system is a little smaller than the Rayleigh wave speed of poroelastic soil. This phenomenon is due to the dynamic interaction between the plate and poroelastic half-space. It is clearly observed in Fig. 2.2.3 that the peak values of the normalized maximum displacement curves decrease as the plate flexural rigidity D_p^* increases.

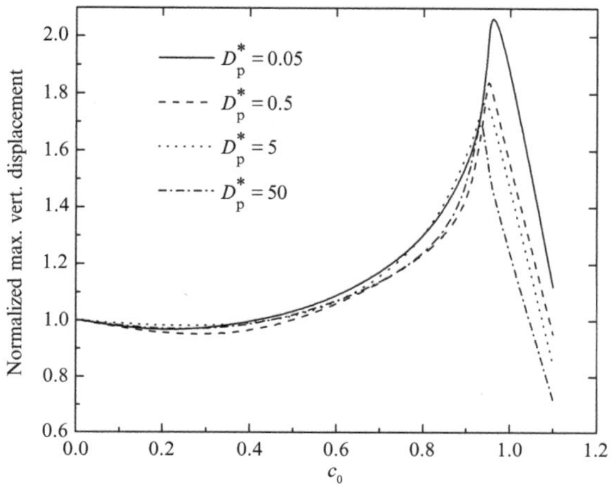

Fig. 2.2.3 Maximum vertical displacement against dimensionless velocity c_0

The acceleration responses at the ground surface are investigated as shown in Fig. 2.2.4. For c_0=0.2 in Fig. 2.2.4A, the acceleration curve has one peak at the center of the tandem axle. The permeability of the soil has little effect on ground acceleration responses. When the load velocity exceeds the Rayleigh wave speed, as shown in Fig. 2.2.4B, the accelerations apparently vibrate to a wide extent, and the magnitude of the accelerations is much larger than that for c_0=0.2. Thus, as the vehicle speed exceeds the critical speed, the vibrations of the road increase significantly. The permeability affects the acceleration responses significantly at this vehicle speed. The acceleration responses decrease as b^* increases.

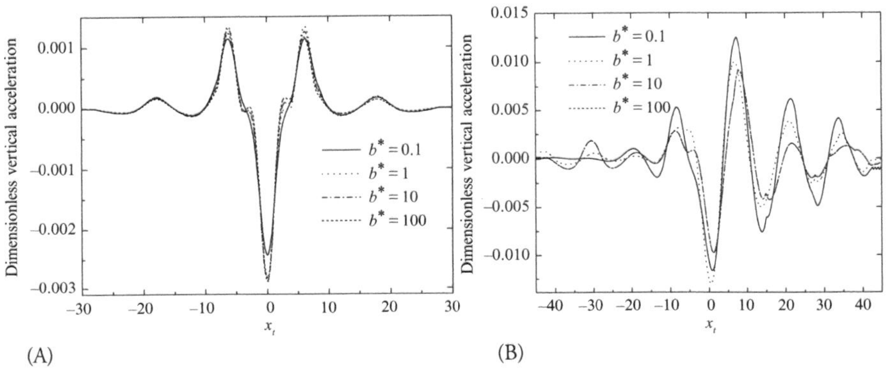

Fig. 2.2.4 Acceleration responses at the surface of the ground. (A) c_0=0.2; (B) c_0=1.0

The vertical stresses of the plate generated by the moving traffic were obtained from the compatibility condition between the plate and the poroelastic half-space. In Fig. 2.2.5, the vertical stress at the interface between the plate and the half-space is plotted against x_t for different D_p^* when c_0=0.2. Clearly, the magnitude of the stress decreases rapidly with increasing D_p^*. When D_p^*=0.05, the positive peak stresses occurred at the point at which the load was applied. And negative stress was caused at the center of the tandem axles because the plate was arched during the load area when the flexural rigidity of the plate was small. As D_p^* increased, the maximum stresses occurred at the center of the tandem axles, and no negative stress was caused. In Fig. 2.2.5B, the vertical stress for high load velocity c_0=1.0 is presented. The vertical stresses of the plate fluctuate significantly faced with the load and the frequency of fluctuations decreased with increasing D_p^*. Compared with Fig. 2.2.5A, the magnitude of the vertical stress increased slightly when the load velocity increased to c_0=1.0. The effect of c_0 on the magnitude of the vertical stress at the contact surface was not pronounced.

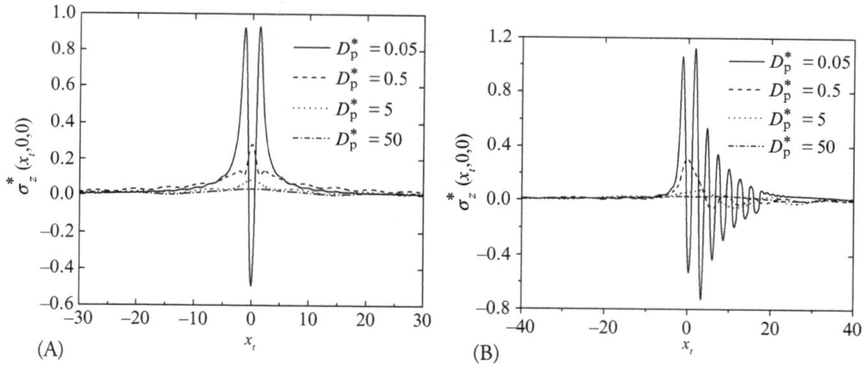

Fig. 2.2.5 The vertical stresses of the plate for different D_p^* . (A) c_0=0.2; (B) c_0=1.0

2.3 Applications in Railway Engineering

In terms of environmental pollution, the consequences of railway traffic become more important in view of the fact that the speed of the trains has increased rapidly, and the vibrations and dynamic stresses in the track base have intensified. It will be important to predict the dynamic responses of the track and ground induced by the moving traffic load.

2.3.1 Governing Equations for the Track System

The vibrations of a saturated poroelastic half-space generated by high-speed trains were first investigated by Cao et al.[7] The vehicle-track-ground coupling system is shown in Fig. 2.3.1. The track model proposed by Picoux and Le Houédec[8] is introduced in this section.

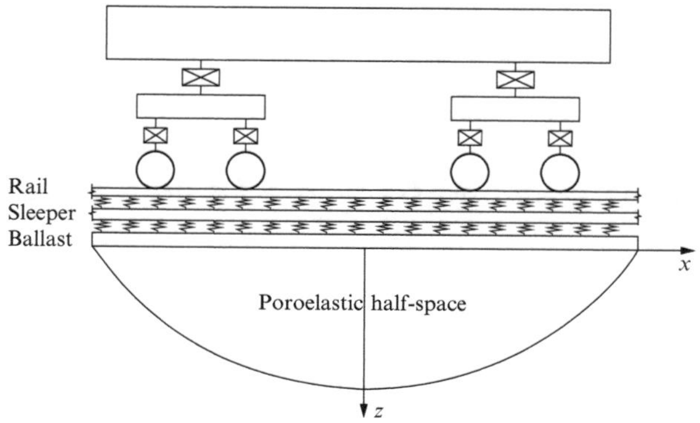

Fig. 2.3.1 Theoretical model of the vehicle-track-ground coupling system

In order to calculate the receptance of the track-ground system, a unit vertical harmonic load $e^{i\Omega t}$

was applied on the rail, which was pointing downwards and located at x=0 when t=0, moving along the rails at speed c. The governing equation for a rail represented by an Euler beam is written as

$$EI\frac{\partial^4 u_\mathrm{R}(x,t)}{\partial x^4}+m_\mathrm{R}\frac{\partial^2 u_R}{\partial t^2}+k_\mathrm{P}[u_\mathrm{R}(x,t)-u_\mathrm{S}(x,t)]=\mathrm{e}^{\mathrm{i}\Omega t}\delta(x-ct) \tag{2.3.1}$$

where u_R is the vertical displacement of the Euler beam, EI is the bending stiffness of the rail beam, m_R is the mass of the rail per unit length, k_P denotes the spring constant of the rail pads, and u_S is the vertical displacement of the sleepers.

It is noted by Knothe and Grassie[9] and Grassie et al.[10] that the effect of the discreteness of the rail supports may be neglected when the dominating frequency content of the response is not in the vicinity of the so-called pinned-pinned resonance frequency. This frequency is normally located in the range 700–1000 Hz, depending on the rail properties and the sleeper distance. The main frequency range of interest for the perception of ground vibration is about 5–80 Hz; therefore it is suitable to represent the sleepers by a continuous mass:

$$m_\mathrm{S}\frac{\partial^2 u_\mathrm{S}(x,t)}{\partial t^2}+k_\mathrm{P}[u_\mathrm{S}(x,t)-u_\mathrm{R}(x,t)]=-F_\mathrm{S}(x,t) \tag{2.3.2}$$

where m_S is the mass of the sleeper per unit length and F_S is the load between the sleepers and the ballast.

The ballast was first considered by the Cosserat model[11]. At the top and bottom of the ballast, the system can be written as

$$\frac{m_\mathrm{B}}{6}\left[2\frac{\partial^2 u_\mathrm{S}(x,t)}{\partial t^2}+\frac{\partial^2 u_\mathrm{B}(x,t)}{\partial t^2}\right]+k_\mathrm{B}[u_\mathrm{S}(x,t)-u_\mathrm{R}(x,t)]=F_\mathrm{S}(x,t) \tag{2.3.3}$$

$$\frac{m_\mathrm{B}}{6}\left[\frac{\partial^2 u_\mathrm{S}(x,t)}{\partial t^2}+2\frac{\partial^2 u_\mathrm{B}(x,t)}{\partial t^2}\right]+k_\mathrm{B}[-u_\mathrm{S}(x,t)+u_\mathrm{B}(x,t)]=-F_\mathrm{B}(x,t) \tag{2.3.4}$$

where m_B is the mass of the ballast per unit length, k_B is the spring constant between ballast and sleepers, F_B is the ballast load on the soil, and u_B is the vertical displacement of the ballast.

2.3.2 Coupling Between the Track and Saturated Ground

The dimensionless variables were adopted to obtain the general trends. All the displacements were nondimensionalized with respect to the unit length a. Pore water pressures and stresses were nondimensionalized with respect to the shear modulus μ. The load between the rail, sleeper, ballast, and ground was nondimensionalized with respect to μa^2. All variables were then replaced by the corresponding dimensionless quantities, denoted by a superscript asterisk (*). The dimensionless time is defined as

$$\tau=(t/a)\sqrt{\mu/\rho} \tag{2.3.5}$$

The following nondimensional parameters are also defined: $k_B^* = \frac{k_B}{\mu a}$, $k_P^* = \frac{k_P}{\mu a}$, $\boldsymbol{K}_V^* = \frac{\boldsymbol{K}_V}{\mu a}$, $\boldsymbol{M}_V^* = \frac{\boldsymbol{M}_V}{\rho a^3}$, $\lambda^* = \frac{\lambda}{\mu}$, $M^* = \frac{M}{\mu}$, $\rho_f^* = \frac{\rho_f}{\rho}$, $m^* = \frac{m}{\rho}$, $b^* = \frac{ab}{\sqrt{\rho\mu}}$, $\beta = \frac{EI}{\mu a^2}$, $m_R^* = \frac{m_R^*}{\rho a^2}$, $m_S^* = \frac{m_S^*}{\rho a^2}$, $c_0 = V/V_s$, V_s is the shear wave velocity of the half-space, expressed as $V_s = \sqrt{\mu/\rho_s}$, a is the unit length, V is the vehicle speed.

The drainage conditions between the track and the poroelastic half-space were assumed to be permeable. Then the boundary conditions of the half-space are given as follows:

$$\tau_{xz}(x^*, y^*, 0, \tau) = 0 \tag{2.3.6}$$

$$\sigma_{zz}(x^*, y^*, 0, \tau) = -\frac{1}{2a}\Pi(y^*)F_B^*(x^*, \tau) \tag{2.3.7}$$

$$\tau_{yz}(x^*, y^*, 0, \tau) = 0 \tag{2.3.8}$$

$$p(x^*, y^*, 0, \tau) = 0 \tag{2.3.9}$$

$$u_z(x^*, 0, 0, \tau) = u_B(x^*, \tau) \tag{2.3.10}$$

The load distributing function is defined as

$$\Pi(y) = \begin{cases} 1, & |y| \le L_{Bal}^* \\ 0, & |y| > L_{Bal}^* \end{cases} \tag{2.3.11}$$

where L_{Bal} is the half-width of the ballast.

From Eqs. (2.3.6) to (2.3.10), the following equation can be obtained:

$$\overline{\overline{\tilde{u}}}_z(\varepsilon, \eta, z^*, \omega) = -\frac{\overline{\Pi}(\eta)}{2L_{Bal}^*}\overline{\overline{\tilde{F}}}{}_B^*(\varepsilon, \omega)\phi(\varepsilon, \eta, z^*, \omega) \tag{2.3.12}$$

where

$$\phi(\varepsilon, \eta, z^*, \Omega) = \frac{1}{\Gamma}\left[\left(\gamma_3{}^2 + \varepsilon^2 + \eta^2\right)\left(\gamma_2 a_2 e^{-\gamma_2 z^*} - \gamma_1 a_1 e^{-\gamma_1 z^*}\right) + (\gamma_1 g_5 - \gamma_2 g_6)e^{-\gamma_3 z^*}\right]$$

$$g_i = E_i + \left(2\varepsilon^2 - b_i{}^2\right)F_i \quad i = 1, 2, \quad g_3 = \lambda^* E_1 - 2\gamma_1^2 F_1 - \alpha, \quad g_4 = \lambda^* E_2 - 2\gamma_2^2 F_2 - \alpha$$

$$g_5 = g_1 + 2\eta^2 F_1, \quad g_6 = g_2 + 2\eta^2 F_2, \quad \Gamma = (g_4 - g_3)\left(\gamma_3^2 + \xi^2 + \eta^2\right) - 2\gamma_3(\gamma_1 g_5 - \gamma_2 g_6)$$

In Eq. (2.3.12), $\overline{\overline{F}}_B$ remains unknown and will be resolved in the next section.

2.3.3 Solutions for the Track-Ground System

Applying the Fourier transform with respect to x^* and τ for Eqs. (2.3.1)–(2.3.4), (2.3.7), the following dimensionless equations can be obtained by eliminating $F_S^*(\xi, \omega)$:

$$\alpha_1(\xi,\omega)\bar{\tilde{u}}_{\mathrm{R}}(\xi,\omega)-k_{\mathrm{P}}^{*}\bar{\tilde{u}}_{\mathrm{S}}(\xi,\omega)=\alpha_2(\xi,\omega) \tag{2.3.13}$$

$$-k_{\mathrm{P}}^{*}\bar{\tilde{u}}_{\mathrm{R}}(\xi,\omega)+\alpha_3(\xi,\omega)\bar{\tilde{u}}_{\mathrm{S}}(\xi,\omega)+\alpha_4(\xi,\omega)\bar{\tilde{u}}_{\mathrm{B}}(\xi,\omega)=0 \tag{2.3.14}$$

$$\alpha_4(\xi,\omega)\bar{\tilde{u}}_{\mathrm{S}}(\xi,\omega)+\alpha_5(\xi,\omega)\bar{\tilde{u}}_{\mathrm{B}}(\xi,\omega)=-\bar{\tilde{F}}_{\mathrm{B}}^{*}(\xi,\omega) \tag{2.3.15}$$

$$\bar{\tilde{u}}_{\mathrm{B}}(\xi,\omega)=\alpha_6(\xi,\omega)\bar{\tilde{F}}_{\mathrm{B}}^{*}(\xi,\omega) \tag{2.3.16}$$

where

$$\alpha_1(\xi,\omega)=\delta\xi^4-m_{\mathrm{R}}^{*}\omega^2+k_{\mathrm{P}}^{*},\quad \alpha_2(\xi,\omega)=2\pi\delta(\omega+\xi c_0-\Omega)$$

$$\alpha_3(\xi,\omega)=-m_{\mathrm{B}}^{*}\omega^2/2+k_{\mathrm{P}}^{*}+k_{\mathrm{B}}^{*}-m_{\mathrm{S}}^{*}\omega^2,\quad \alpha_4(\xi,\omega)=-m_{\mathrm{B}}^{*}\omega^2/6-k_{\mathrm{B}}^{*}$$

$$\alpha_5(\xi,\omega)=-m_{\mathrm{B}}^{*}\omega^2/3+k_{\mathrm{B}}^{*},\quad \alpha_6(\xi,\omega)=\Psi(\xi,0,\omega)$$

$$\Psi(\xi,0,\omega)=\frac{1}{2\pi}\int_{-\infty}^{\infty}-\bar{\Pi}(\eta)\phi(\xi,\eta,0,\omega)\mathrm{e}^{\mathrm{i}\eta y}\,\mathrm{d}y$$

By using Eqs. (2.3.13)–(2.3.16), $\bar{\tilde{F}}_{\mathrm{B}}(\xi,\omega)$ is obtained:

$$\bar{\tilde{F}}_{\mathrm{B}}^{*}(\xi,\omega)=\frac{\alpha_2(\xi,\omega)\alpha_4(\xi,\omega)k_{\mathrm{P}}^{*}}{\alpha_1(\xi,\omega)\alpha_6(\xi,\omega)\alpha_4{}^2(\xi,\omega)-\left(\alpha_3(\xi,\omega)\alpha_1(\xi,\omega)-k_{\mathrm{P}}^{*}2\right)\left(1+\alpha_5(\xi,\omega)\alpha_6(\xi,\omega)\right)} \tag{2.3.17}$$

Thus, the rail displacement in the Fourier transform domain is finally given by

$$\bar{\tilde{u}}_{\mathrm{R}}^{*}(\xi,\omega)=\frac{-\left(\alpha_2(\xi,\omega)\alpha_4(\xi,\omega)^2-\alpha_2(\xi,\omega)\alpha_3(\xi,\omega)\alpha_5(\xi,\omega)-\alpha_4(\xi,\omega)\bar{\tilde{F}}_{\mathrm{B}}^{*}(\xi,\omega)k_{\mathrm{p}}^{*}\right)}{\left(\alpha_1(\xi,\omega)\alpha_4(\xi,\omega)^2-\alpha_1(\xi,\omega)\alpha_3(\xi,\omega)\alpha_5(\xi,\omega)+\alpha_5(\xi,\omega)k_{\mathrm{p}}^{*}2\right)} \tag{2.3.18}$$

The displacement of the rail and ground in the time domain can be expressed as follows by introducing an auxiliary spatial coordinate $x_t{}^{*}=x^{*}-c_0\tau$, and the time-domain results are obtained by the FFT algorithm.

$$u_{\mathrm{R}}^{*}(x^{*},\tau)=\frac{1}{2\pi}\int_{-\infty}^{\infty}\bar{\tilde{u}}_{\mathrm{R}}^{*}\;(\xi,\Omega^{*}-\xi c_0)\mathrm{e}^{\mathrm{i}\xi x_t^{*}}\,\mathrm{d}\xi\cdot\mathrm{e}^{\mathrm{i}\Omega^{*}\tau} \tag{2.3.19}$$

$$u_z^{*}(x^{*},y^{*},z^{*},\tau)=\frac{1}{4\pi^2}\int_{-\infty}^{\infty}\int_{-\infty}^{\infty}\bar{\tilde{u}}_z^{*}\;(\xi,\eta,\Omega^{*}-\xi c_0)\mathrm{e}^{\mathrm{i}\eta y^{*}}\mathrm{e}^{\mathrm{i}\xi x_t^{*}}\,\mathrm{d}\eta\,\mathrm{d}\xi\cdot\mathrm{e}^{\mathrm{i}\Omega\;\tau} \tag{2.3.20}$$

Eqs. (2.3.19), (2.3.20) can also be expressed as

$$u_{\mathrm{R}}^{*}(x,\tau)=u_{\mathrm{R}}^{\Omega}\;(x_t{}^{*})\cdot\mathrm{e}^{\mathrm{i}\Omega^{*}\tau} \tag{2.3.21}$$

$$u_z^*(x^*, y^*, z^*, \tau)=u_z^{\Omega}(x_\tau^*, y^*, z^*)\cdot e^{i\Omega^*\tau} \quad (2.3.22)$$

Eqs. (2.3.21), (2.3.22) denote that, in the auxiliary spatial coordinate, the displacements of the track-ground system are harmonic and have the same vibration frequency as the dynamic load.

2.3.4 Coupling of Vehicle-Track-Ground System

The receptance herein denotes the displacement amplitude of the wheelsets due to a unit vertical harmonic load with an excitation frequency Ω. The vehicle model used in Sheng and Jones[12] was introduced. As shown in Fig. 2.3.1, the vehicles were represented as multiple rigid body systems and the vertical dynamics of the vehicles were coupled to the track-ground model by introducing linear Hertzian contact springs between each wheelset and the rails. The differential equation of motion for a single vehicle is given by

$$\boldsymbol{M}_{\mathrm{V}}\ddot{\boldsymbol{Z}}_{\mathrm{V}}(t)+\boldsymbol{K}_{\mathrm{V}}\boldsymbol{Z}_{\mathrm{V}}(t)=-\boldsymbol{BP}(t) \quad (2.3.23)$$

where $\boldsymbol{M}_{\mathrm{V}}$ and $\boldsymbol{K}_{\mathrm{V}}$ denote the mass and stiffness matrices of the vehicle respectively, $\boldsymbol{Z}_{\mathrm{V}}(t)$ denotes the displacement vector, $\boldsymbol{P}(t)$ denotes the wheel-rail force vector, and $\boldsymbol{B}$ is a matrix of unit and zero elements (see Appendix A).

The roughness-induced dynamic loads between the wheel and rail were harmonic loads with angular frequency Ω, where $\Omega=2\pi c/\lambda_1$, λ_1 is the wavelength of the rail profile, and c is the vehicle speed. As shown by Sheng and Jones[12], $\boldsymbol{P}(t)$ and $\boldsymbol{Z}_{\mathrm{V}}(t)$ can be expressed as $\boldsymbol{P}(t)=\boldsymbol{P}'(\Omega)\,e^{i\Omega t}$ and $\boldsymbol{Z}_{\mathrm{V}}(t)=\boldsymbol{Z}'_{\mathrm{V}}(\Omega)\,e^{i\Omega t}$. Then Eq. (2.3.23) can be written as

$$\boldsymbol{Z}'_{\mathrm{V}}(\Omega)=-(\boldsymbol{K}_{\mathrm{V}}-\Omega^2\boldsymbol{M}_{\mathrm{V}})^{-1}\boldsymbol{BP}'(\Omega) \quad (2.3.24)$$

The receptance between the jth and kth wheelsets within a vehicle is denoted by Δ_{jk}^{W} (W means wheelset), where j, k = 1, 2, …, N; N is the number of wheelsets of the vehicle. Δ_{jk}^{W} denotes the displacement amplitude of the jth wheelset due to a unit vertical harmonic load with an excitation frequency Ω exerted at the kth wheelset. The displacement vector of the wheelset in the vehicle is expressed as

$$\boldsymbol{Z}'_{\mathrm{W}}(\Omega)=(Z'_{\mathrm{W}1}(\Omega), Z'_{\mathrm{W}2}(\Omega), Z'_{\mathrm{W}3}(\Omega),\ldots,Z'_{\mathrm{W}N}(\Omega))^{\mathrm{T}} \quad (2.3.25)$$

and

$$\boldsymbol{P}'(\Omega)=(P'_1(\Omega), P'_2(\Omega), P'_3(\Omega),\ldots, P'_N(\Omega))^{\mathrm{T}} \quad (2.3.26)$$

is the wheel-rail force vector for a vehicle.

The displacement vector of the wheelsets is part of that for the corresponding vehicle. Therefore, it can be written as

$$\boldsymbol{Z}'_{\mathrm{W}}(\Omega)=\boldsymbol{A}\boldsymbol{Z}'_{\mathrm{V}}(\Omega) \tag{2.3.27}$$

where $\boldsymbol{A}=\boldsymbol{B}^{\mathrm{T}}$ (see Appendix A). Thus

$$\boldsymbol{Z}'_{\mathrm{W}}(\Omega)=-\Delta^{\mathrm{W}}\boldsymbol{P}'(\Omega)=-\mathrm{A}(\boldsymbol{K}_{\mathrm{V}}-\Omega^2\boldsymbol{M}_{\mathrm{V}})^{-1}\boldsymbol{B}\boldsymbol{P}'(\Omega) \tag{2.3.28}$$

$$\Delta^{\mathrm{W}}=\begin{bmatrix}\Delta_{11}^{\mathrm{W}} & \cdots & \Delta_{1N}^{\mathrm{W}}\\ \vdots & \cdots & \vdots\\ \Delta_{N1}^{\mathrm{W}} & \cdots & \Delta_{NN}^{\mathrm{W}}\end{bmatrix}=\boldsymbol{A}\left(\boldsymbol{K}_{\mathrm{V}}-\Omega^2\boldsymbol{M}_{\mathrm{V}}\right)^{-1}\boldsymbol{B} \tag{2.3.29}$$

Eq. (2.3.29) gives the receptance matrix at the wheelsets for a single vehicle. Suppose there are N_1 identical vehicles being considered; the total number of the wheel-rail loads is $M=N_1N$. The receptance matrix at the wheelsets for the train, denoted by Δ^{T} (where T stands for train), is given by

$$\Delta^{\mathrm{T}}=\mathrm{diag}\left(\Delta^{\mathrm{W}},\ \ldots,\ \Delta^{\mathrm{W}}\right)=\begin{bmatrix}\Delta^{\mathrm{W}} & \cdots & 0\\ \vdots & \vdots & \vdots\\ 0 & \cdots & \Delta^{\mathrm{W}}\end{bmatrix} \tag{2.3.30}$$

The elements of matrix Δ^{T} are denoted by $\Delta_{lk}{}^{\mathrm{T}}$, where k, l = 1, 2, …, M.

Thus, the receptance at the jth wheel-rail contact point due to a unit load at the kth wheel-rail contact point on the rail is determined by

$$\Delta_{jk}^{\mathrm{R}}=u_{\mathrm{R}}^{\Omega}(l_{jk}^{*}) \tag{2.3.31}$$

where

$$l_{jk}^{*}=a_j^{*}-a_k^{*} \tag{2.3.32}$$

is the dimensionless distance between the two contact points.

The complex amplitudes of the displacements at the wheel-rail contact points on the rails are given by

$$\boldsymbol{z}'^{*}_{\mathrm{R}}(\Omega^{*})=\Delta^{\mathrm{R}}\boldsymbol{P}^{*}(\Omega^{*}) \tag{2.3.33}$$

where

$$\Delta^{\mathrm{R}}=\begin{bmatrix}\Delta_{11}^{\mathrm{R}} & \Delta_{12}^{\mathrm{R}} & \cdots & \Delta_{1M}^{\mathrm{R}}\\ \Delta_{21}^{\mathrm{R}} & \Delta_{22}^{\mathrm{R}} & \cdots & \Delta_{2M}^{\mathrm{R}}\\ \vdots & \vdots & \vdots & \vdots\\ \Delta_{M1}^{\mathrm{R}} & \Delta_{M2}^{\mathrm{R}} & \cdots & \Delta_{MM}^{\mathrm{R}}\end{bmatrix} \tag{2.3.34}$$

The receptance matrix is nonsymmetrical due to the load motion.

$$\boldsymbol{z}_{\mathrm{R}}^{\prime *}(\Omega^*)=(z_{\mathrm{R1}}^{\prime *}(\Omega^*), z_{\mathrm{R2}}^{\prime *}(\Omega^*), z_{\mathrm{R3}}^{\prime *}(\Omega^*), \ldots, z_{\mathrm{RM}}^{\prime *}(\Omega^*))^{\mathrm{T}} \tag{2.3.35}$$

Eq. (2.3.35) represents the displacement vector of the rail at the wheel-rail contact points observed in the auxiliary spatial coordinate.

The rail irregularities are presented by a sinusoidal profile of amplitude A. The profile of the rail irregularities is given by

$$z(x)=A\mathrm{e}^{\mathrm{i}(2\pi/\lambda_1)x} \tag{2.3.36}$$

where λ_1 denotes the wavelength. The process was assumed to be linear, so that a displacement input was generated at the excitation frequency $f=c/\lambda_1$. The angular frequency is obtained by $\Omega=2\pi c/\lambda_1$. At time t, the lth wheelset arrives at $x=a_l+ct$; thus the displacement input of the rail profile at the lth wheel-rail contact point is

$$z_l(\mathrm{t})=z'_l(\Omega)\,\mathrm{e}^{\mathrm{i}\Omega t}=A\mathrm{e}^{\mathrm{i}(2\pi/\lambda_1)(a_l+ct)}=A\mathrm{e}^{\mathrm{i}(2\pi/\lambda_1)a_l}\,\mathrm{e}^{\mathrm{i}\Omega t} \tag{2.3.37}$$

The coupling of a wheelset with rail is illustrated in Fig. 2.3.2, where $z'_{\mathrm{W}l}(\Omega)\mathrm{e}^{\mathrm{i}\Omega t}$ denotes the displacement of the lth wheelset.

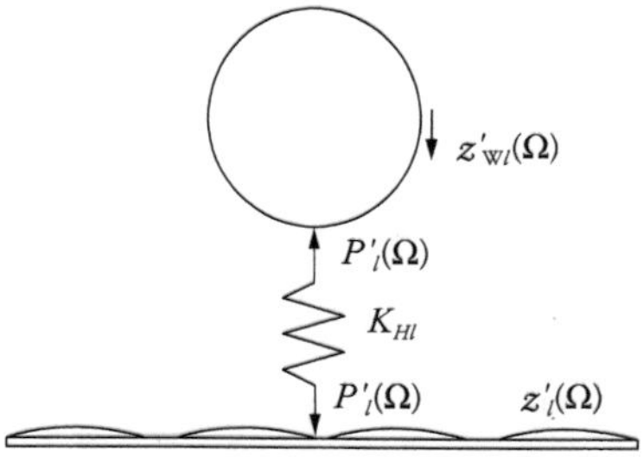

Fig. 2.3.2 Coupling wheel-rail model

The wheel and rail deform locally according to the Hertz theory under the action of the contact force. Thus the wheel and rail are coupled by a Hertz spring. Provided that the dynamic contact force is a small fraction of the axle load, the Hertz spring can be taken to be linear. The stiffness of the Hertzian contact spring is denoted by k_{Hl}. It is also assumed that the wheelset is always in contact with the rail. Thus

$$z'_{\mathrm{W}l}(\Omega)=z'_{\mathrm{R}l}(\Omega)+z'_l(\Omega)+P'(\Omega)/k_{Hl} \tag{2.3.38}$$

From Eqs. (2.3.21) to (2.3.28), the following relations can be derived:

$$z'_{\mathrm{W}l}(\Omega) = -\sum_{k=1}^{M} \Delta_{lk}^{\mathrm{T}} P'_k(\Omega) \tag{2.3.39}$$

$$z'_{\mathrm{R}l}(\Omega) = \sum_{k=1}^{M} \Delta_{lk}^{\mathrm{R}} P'_k(\Omega) \tag{2.3.40}$$

By nondimensionalizing these previous equations and substituting Eqs. (2.3.39), (2.3.40) into Eq. (2.3.38) the following equation can be obtained

$$\sum_{k=1}^{M} (\Delta_{lk}^{\mathrm{T}} + \Delta_{lk}^{\mathrm{R}}) P'^{*}_k(\Omega^*) + P'^{*}_l(\Omega^*)/k^{*}_{Hl} = -z'^{*}_l(\Omega^*) \tag{2.3.41}$$

The unknown item $P'_l(\Omega^*)$ can be obtained by solving Eq. (2.3.41). The displacement of the ground and the rails at the exciting frequency Ω^* are given by superposition.

$$u^{*}_{\mathrm{R}}(x^*, \tau) = \sum_{l=1}^{M} u^{\Omega}_{\mathrm{R}}(x_{\mathrm{t}}{}^* - a^*{}_l) P'_l(\Omega^*) \mathrm{e}^{\mathrm{i}\Omega^*\tau} \tag{2.3.42}$$

$$u^{*}_{\mathrm{z}}(x^*, \tau) = \sum_{l=1}^{M} u^{\Omega}_{\mathrm{z}}(x_{\mathrm{t}}{}^* - a_l{}^*, y^*, z^*) P'_l(\Omega^*) \mathrm{e}^{\mathrm{i}\Omega^*\tau} \tag{2.3.43}$$

2.3.5 Numerical Results

In order to demonstrate the characteristics of the response of the system, the calculations were carried out for a single-axle vehicle model comprising a suspended mass M_{C} and an unsprung mass M_{W}. A detailed description of the model is given in Appendix A. Each mass had one degree of freedom in the vertical direction. The frequency of interest for the perception of ground vibration was about 5–80 Hz, and for a train speed in the range of 10–120 m/s, the corresponding wavelengths of the corrugated rail lay within the range of 0.125–24 m. The parameters for the vehicle model are listed in Table 2.3.1 and the parameters for the track are given in Table 2.3.2, and correspond to those used in Sheng and Jones[12]. The parameters of the saturated poroelastic half-space are presented in Table 2.3.3, which are the same as those used in Cai et al.[4] The parameters were selected according to Tables 2.3.1–2.3.3 if not denoted in the figures.

Table 2.3.1 Dimensionless parameters for the single-axle vehicle model

Parameter	Value
Suspended mass, $M_{\mathrm{C}}{}^*$	12.9
Unsprung mass, $M_{\mathrm{W}}{}^*$	1.173
$k_{\mathrm{S}}{}^*$	0.089
$C_{\mathrm{S1}}{}^*$	0.165
k_{S}^{*}	0.1

Table 2.3.2 Dimensionless parameters for railway track

Parameter	Value
Mass of the beam per unit length of track, m_R^*	0.08
Bending stiffness of the rail beam, β	0.42
Loss factor of the rail	0.01
Rail pad stiffness, k_P^*	11.67
Rail pad loss factor	0.15
Mass of the sleeper per unit length of the track, m_S^*	0.328
Mass of the ballast per unit length of the track, m_B^*	0.84
Ballast stiffness per unit length of the track , k_B^*	10.5
Loss factor of the ballast	1.0
Contact width of ballast and ground, $2L_{Bal}^*$	2.7

Table 2.3.3 Dimensionless parameters for fully water-saturated poroelastic soil medium

Parameter	Value
Lamé constant, λ^*	2
Water density, $\rho_f{}^*$	0.53
Parameter of soil structure, m^*	1.5625
Hysteretic damping ratio, D	0.02
Ratio between the fluid viscosity and the intrinsic permeability, b^*	10
The parameter for the compressibility of the soil particle, α	0.97
The parameter for the compressibility of the fluid, M^*	12

The roughness-induced dynamic wheel-rail forces are shown in Fig. 2.3.3 against the excitation frequency f^* ($f^*=c_0/\lambda_1^*$). It is shown that the magnitude of the wheel-rail forces increased as f^* increased and reached a maximum value at the critical frequency around , then decreased as f^* increased further. With a close inspection of Fig. 2.3.3, one can also see that the magnitude of the dynamic wheel-rail forces have a local peak at the frequency 0.02. This frequency is close to the natural frequency of the suspended mass on the suspension, as noted in Sheng and Jones[12].

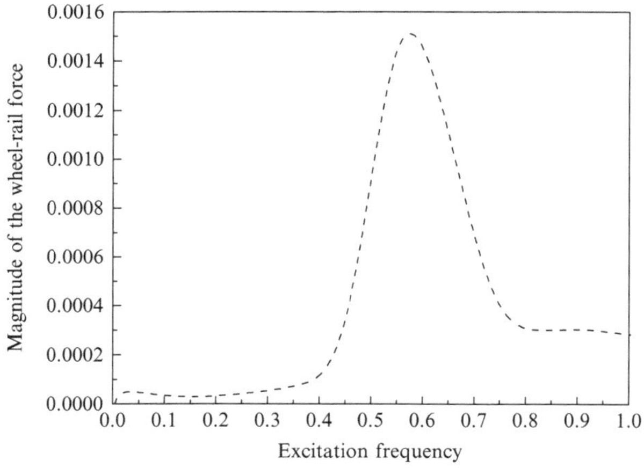

Fig. 2.3.3 Dynamic wheel-rail force at different excitation frequencies

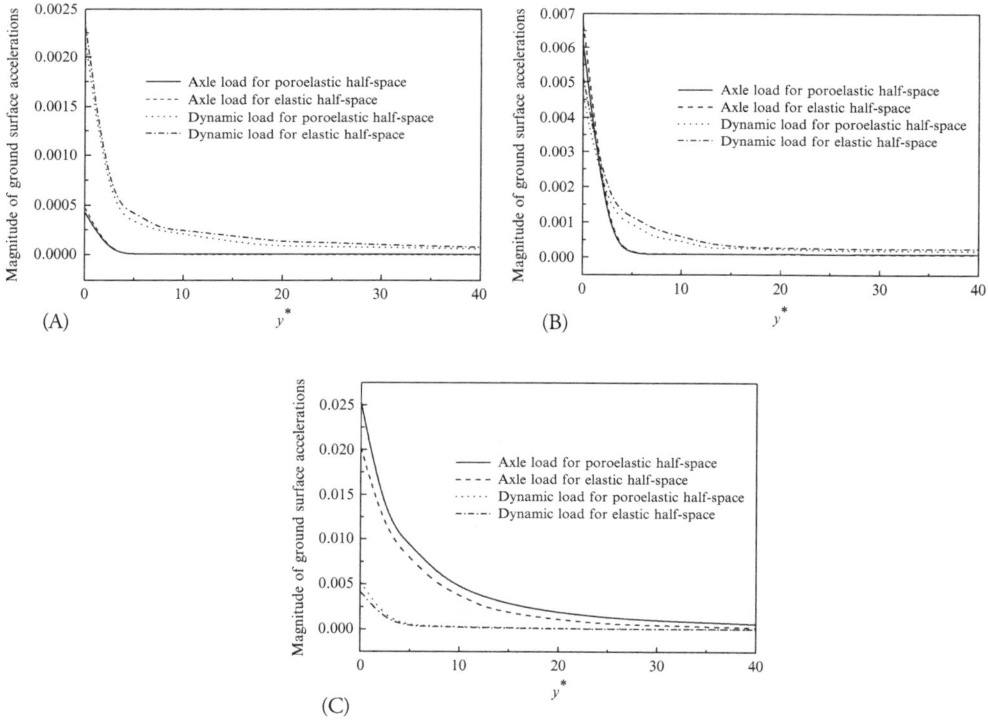

Fig. 2.3.4 Ground surface accelerations against the distance. (A) c_0=0.2; (B) c_0=0.6; (C) c_0=1.0

In order to study the attenuation of the ground vibration, the magnitudes of accelerations for both axle and dynamic loads are presented against y^* in Fig. 2.3.4. In Fig. 2.3.4A, when $c_0=0.2$, the quasistatically induced acceleration decreases rapidly as y^* increases, and is nearly zero at $y^*=5$. The dynamically induced acceleration is much larger at $y^*=0$ and decreases more slowly. The dynamically induced component of vibration is the dominant one for the acceleration response of the free-field. In Fig. 2.3.4A, the acceleration response for the elastic half-space soil medium is also presented. The acceleration responses predicted by the elastic soil model are larger than those predicted by the poroelastic soil model for both axle and dynamic loads at $y^*=0$, and the acceleration responses attenuate more slowly in the elastic soil medium. The free-field acceleration response predicted by the elastic soil model is larger than that predicted by the poroelastic one. In Fig. 2.3.4B, when $c_0=0.6$, the quasistatically induced acceleration is larger than the dynamically induced acceleration at $y^*=0$, but the dynamically induced vibration component dissipates more slowly and gives rise to the majority of the vibration for the free-field off the track. For the case of an elastic soil medium, it can be seen that the acceleration response dissipates more slowly in the elastic soil. Using the elastic soil medium model will overestimate the acceleration response under the free-field condition for $c_0=0.6$. In Fig. 2.3.4C, when $c_0=1.0$, the quasistatically induced accelerations are much larger and are the dominant component for the ground vibration at various y^*. The acceleration response predicted by the elastic soil model is apparently smaller than that predicted by the poroelastic soil model.

Therefore, for vehicle speeds below the Rayleigh wave speed, the free-field acceleration response is dominated by the dynamic load. But when the vehicle speed approaches the Rayleigh wave speed of the half-space, the axle load is the dominant one for the acceleration response of the free-field. Using the elastic soil medium model will overestimate the acceleration response level of the free-field for the train speed below Rayleigh wave speed, while it will underestimate the response level for the train speed above the Rayleigh wave speed.

In Fig. 2.3.5, the train-induced excess pore water pressures are studied. In Fig. 2.3.4A, when $c_0=0.2$, the quasistatically induced pore water pressure reaches a positive peak at the load point and a negative peak behind the load. The dynamically induced pore water pressure fluctuates along x_t^* and the maximum value is larger than that for the quasistatically induced one. When the vehicle speed exceeds the critical speed, as shown in Fig. 2.3.5B, the quasistatically induced pore water pressure becomes much larger and is the dominant component for the pore water pressure responses.

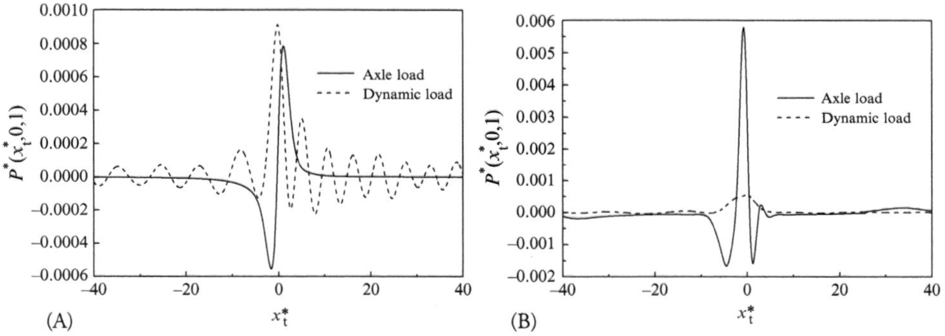

Fig. 2.3.5 Excess pore water pressure responses

2.4 Project Case

The Beijing-Tianjin intercity railway line is the earliest constructed high-speed railway line in China; it connects the two municipalities of Beijing and Tianjing. Construction of this intercity railway line was started on July 4, 2005 and was completed on Dec. 15, 2007. The designed operating speed was 350 km/h. In order to ensure ride quality as well as reduce maintenance costs, the slab track Bögl imported from Germany was adopted on the railway line. The subgrade of the railway lines was mainly composed of soft clay, and then treated by CFG piles with a diameter of 0.4 m and a length of 28 m. After debugging, the railway started test runs on July 1, 2008. During this period, Zhejiang University was invited to measure the dynamic responses of the track and embankment. Two locomotive types, Harmony CRH2 and CRH3, were used during the test runs. The measured vibration of the track and embankment is presented herein.

An in-situ test of the ground-borne vibration induced by the Beijing-Tianjin railway lines was conducted by the Institute of Geotechnical Engineering, Zhejiang University[13]. The measured results was also presented in Song et al.[14], and some of them were given as following.

The testing site was chosen at Yangcun. The instrumentation of the velocity sensor and the accelerometer on the cross-section of the railway line is shown in Fig. 2.4.1. V and H denote the sensors in the vertical and horizontal directions, respectively.

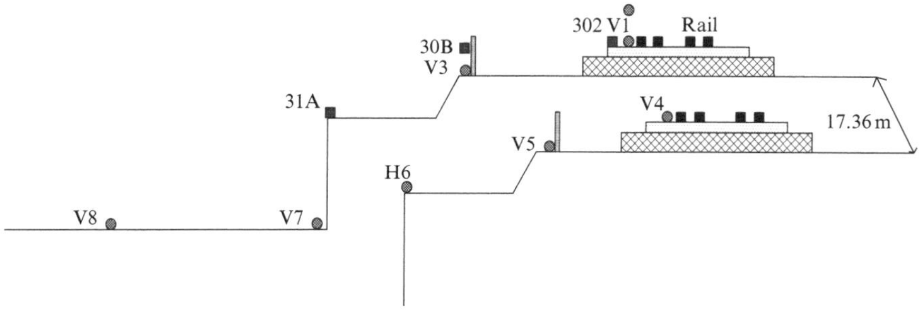

Fig. 2.4.1 Cross-section arrangement of the sensors for the vibration test

The measured results are as follows. When the train runs at the speed of 145.1 km/h, the measured velocity at the sleepers, embankment, and free field are presented in Figs. 2.4.2–2.4.4. The peak velocity at the sleeper was about 1.0 cm/s, and then decreased as the distance from the railway centerline increased. When the train speed increased to 335.1 km/h, the measured velocity at the sleepers, embankment, and free field are presented in Figs. 2.4.5–2.4.7, and it can be seen that the velocity at the sleeper was amplified to 1.5 cm/s, then decreased as the distance increased.

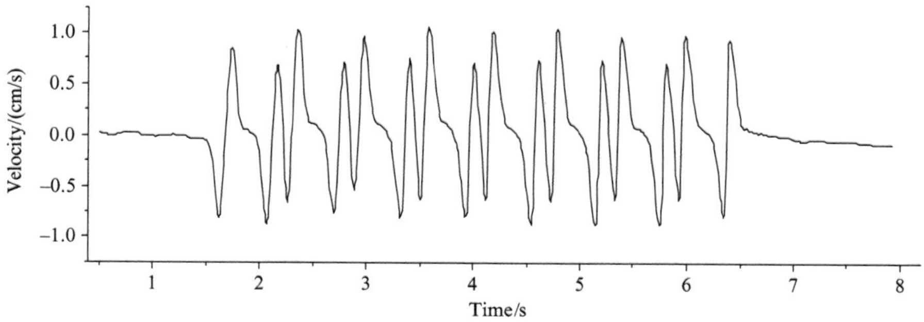

Fig. 2.4.2 Measured velocity at the sleeper (V1) under a train speed of 145.1 km/h (maximum value 1.046cm/s, minimum value - 0.906cm/s)

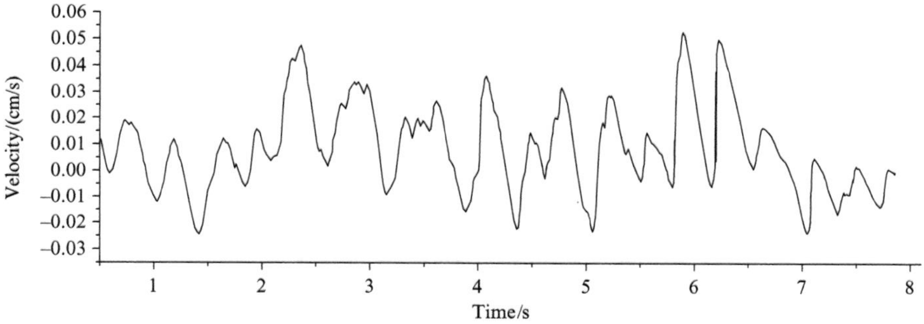

Fig. 2.4.3 Measured velocity at the embankment (V3) under a train speed of 145.1 km/h (maximum value 0.058cm/s, minimum value - 0.025cm/s)

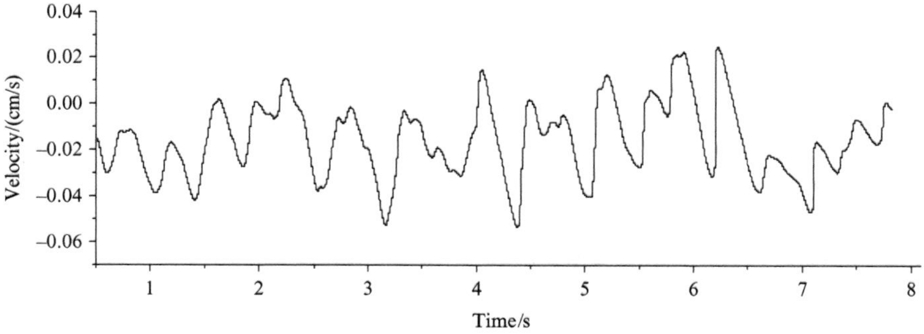

Fig. 2.4.4 Measured velocity at the free-field (V7) under a train speed of 145.1 km/h (maximum value 0.026cm/s, minimum value - 0.053cm/s)

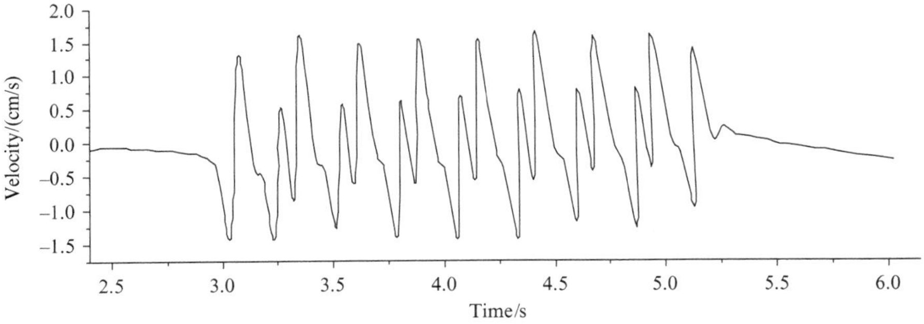

Fig. 2.4.5 Measured velocity at sleeper (V1) under a train speed of 335.1 km/h (maximum value 1.866cm/s, minimum value - 1.419cm/s)

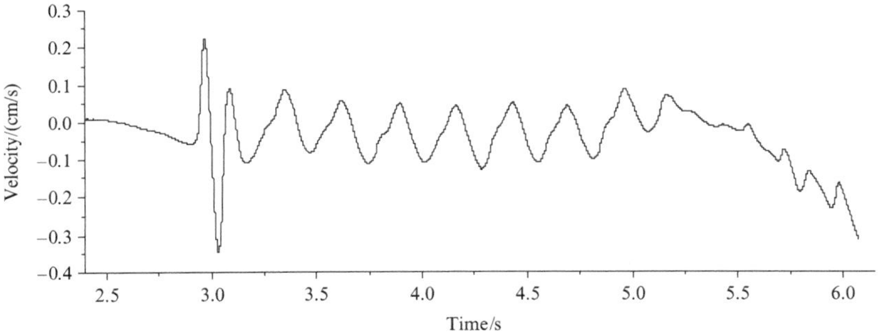

Fig. 2.4.6 Measured velocity at the embankment (V3) under a train speed of 335.1 km/h (maximum value 0.245cm/s, minimum value - 0.348cm/s)

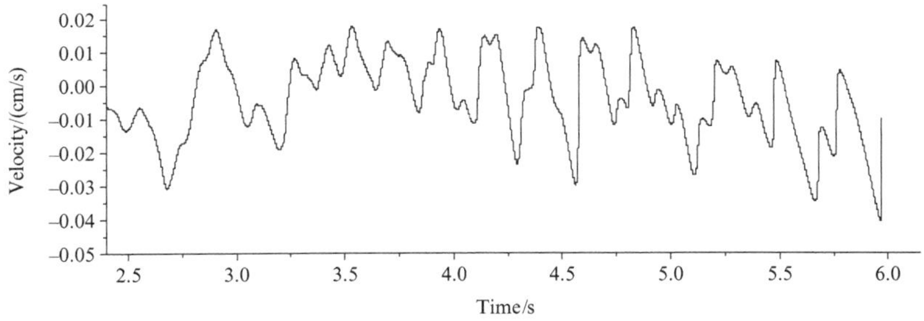

Fig. 2.4.7 Measured velocity at the free-field (V8) under a train speed of 335.1 km/h (maximum value 0.020cm/s, minimum value - 0.031cm/s)

In order to validate the correctness of the theoretical model proposed in the previous section, the comparison between the calculated and measured velocity at the sleeper (V1) under a train speed of 145.1 km/h and 335.1 km/h is shown in Figs. 2.4.8 and 2.4.9, respectively. It is shown that the calculated and measured results agreed well.

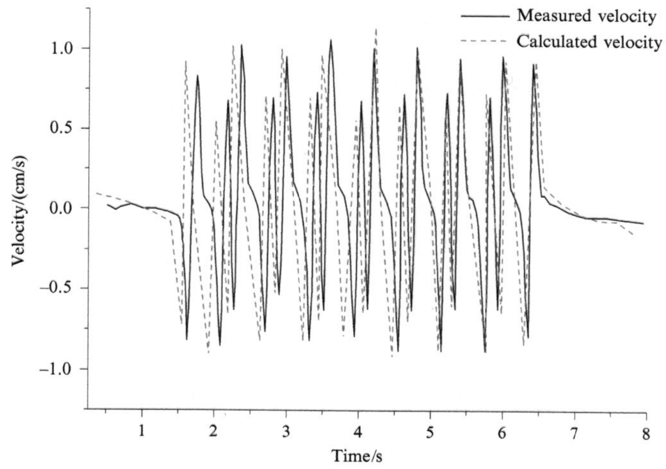

Fig. 2.4.8 The calculated and measured velocities at the sleeper (V1) under a train speed of 145.1 km/h

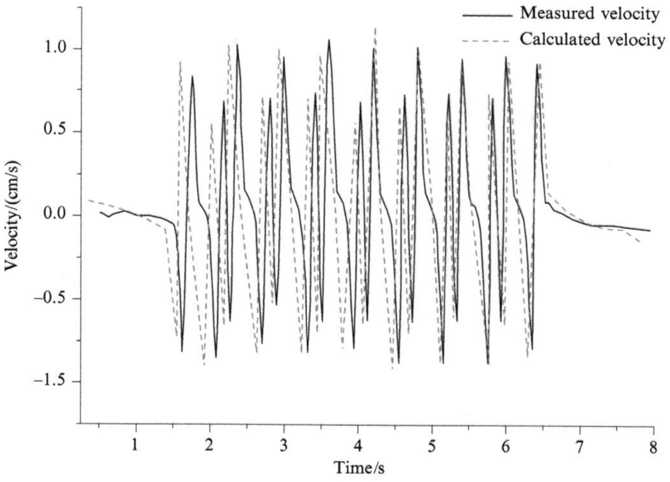

Fig. 2.4.9 The calculated and measured velocities at the sleeper (V1) under a train speed of 335.1 km/h

Based on the in situ test results, the following results were obtained:

(1) The velocity of the track was intensified significantly as the train speed increased; the velocity amplitude of the track at 350 km/h was about two times what it was at 120 km/h.

(2) The vibration decayed rapidly along the transverse direction; the vibration velocity at the free-field was significantly smaller than that at the sleeper of the track.

Chapter 3
Solutions for Vibrations from Underground Railways

This chapter introduces two analytical models for predicting ground vibrations from a tunnel embedded in the saturated poroelastic soil. One model is of a tunnel surrounded by a poroelastic full-space and the other model is of a tunnel buried in a poroelastic half-space. The model of a tunnel embedded in a poroelatic full-space is suitable for the qualitative and general trend analysis as the ground is modeled as a full-space. In the model of a tunnel embedded in a half-space, the ground surface is taken into account. The above two models and their applications in the underground railway engineering will be presented in Sections 3.1 and 3.2, respectively.

3.1 Dynamic Response of a Tunnel Embedded in a Saturated Poroelastic Full-Space

The model of a tunnel embedded in the poroelatic full-space is an extension of the famous Pipe-in-Pipe model proposed by Hunt, Forrest and Hussein in University of Cambridge[14–16]. In order to include the influence of pore fluid in the soil, Biot's poroelastic theory would be used to govern the behavior of two-phase soil medium. The circular tunnel is modelled as a thin shell, as shown in Fig. 3.1.1, with a radius a and a thickness h.

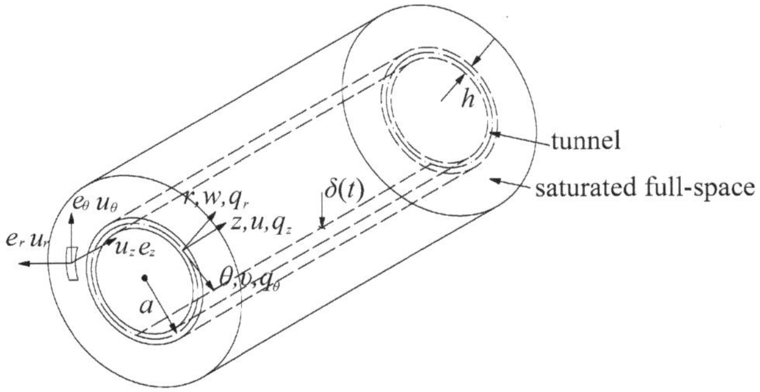

Fig. 3.1.1 A circular tunnel embedded in the saturated poroelastic full-space

3.1.1 Solution for a Cavity in the Saturated Poroelastic Full-Space

In cylindrical coordinates, the governing equations of a fully saturated poroelastic medium are given as Eq. (1.1.28).

The displacement field can be described by scalar and vector potentials

$$\boldsymbol{u}=\nabla\varphi+\nabla\times\boldsymbol{\psi} \tag{3.1.1}$$

$$\boldsymbol{w}=\nabla\chi+\nabla\times\boldsymbol{\Theta} \tag{3.1.2}$$

where φ and χ are scalar potentials of the solid skeleton and the pore fluid, respectively; $\boldsymbol{\psi}$ and $\boldsymbol{\Theta}$ are vector potentials of the solid skeleton and the pore fluid, respectively.

The substitution of wave potentials into Eq. (1.1.28) gives

$$\begin{bmatrix}\lambda+2\mu+\alpha^2M & \alpha M\\ \alpha M & M\end{bmatrix}\begin{bmatrix}\nabla^2\varphi\\ \nabla^2\chi\end{bmatrix}=\begin{bmatrix}\rho\dfrac{\partial^2}{\partial t^2} & \rho_{\mathrm{f}}\dfrac{\partial^2}{\partial t^2}\\ \rho_{\mathrm{f}}\dfrac{\partial^2}{\partial t^2} & m\dfrac{\partial^2}{\partial t^2}+\dfrac{\partial}{\partial t}b\end{bmatrix}\begin{bmatrix}\varphi\\ \chi\end{bmatrix} \tag{3.1.3}$$

$$\begin{bmatrix}\mu & 0\\ 0 & 0\end{bmatrix}\begin{bmatrix}\nabla^2\boldsymbol{\psi}\\ \nabla^2\boldsymbol{\Theta}\end{bmatrix}=\begin{bmatrix}\rho\dfrac{\partial^2}{\partial t^2} & \rho_{\mathrm{f}}\dfrac{\partial^2}{\partial t^2}\\ \rho_{\mathrm{f}}\dfrac{\partial^2}{\partial t^2} & m\dfrac{\partial^2}{\partial t^2}+\dfrac{\partial}{\partial t}b\end{bmatrix}\begin{bmatrix}\boldsymbol{\psi}\\ \boldsymbol{\Theta}\end{bmatrix} \tag{3.1.4}$$

where $\nabla^2=\dfrac{1}{r}\dfrac{\partial}{\partial r}+\dfrac{\partial^2}{\partial r^2}+\dfrac{1}{r^2}\dfrac{\partial^2}{\partial\theta^2}+\dfrac{\partial^2}{\partial z^2}$.

If we transform Eqs. (3.1.3)–(3.1.4) into the frequency domain, they becomes

$$\begin{bmatrix}\lambda+\alpha^2M+2\mu & \alpha M\\ \alpha M & M\end{bmatrix}\begin{bmatrix}\nabla^2\tilde{\varphi}\\ \nabla^2\tilde{\chi}\end{bmatrix}=\begin{bmatrix}-\rho\omega^2 & -\rho_{\mathrm{f}}\omega^2\\ -\rho_{\mathrm{f}}\omega^2 & -m\omega^2+\mathrm{i}\omega b\end{bmatrix}\begin{bmatrix}\tilde{\varphi}\\ \tilde{\chi}\end{bmatrix} \tag{3.1.5}$$

$$\begin{bmatrix}\mu & 0\\ 0 & 0\end{bmatrix}\begin{bmatrix}\nabla^2\tilde{\boldsymbol{\psi}}\\ \nabla^2\tilde{\boldsymbol{\Theta}}\end{bmatrix}=\begin{bmatrix}-\rho\omega^2 & -\rho_{\mathrm{f}}\omega^2\\ -\rho_{\mathrm{f}}\omega^2 & -m\omega^2+\mathrm{i}\omega b\end{bmatrix}\begin{bmatrix}\tilde{\boldsymbol{\psi}}\\ \tilde{\boldsymbol{\Theta}}\end{bmatrix} \tag{3.1.6}$$

Decoupling Eqs. (3.1.5) and (3.1.6), the standard wave equation can be obtained

$$\nabla^2\tilde{\varphi}_{\mathrm{f,s}}+k_{\mathrm{f,s}}^2\tilde{\varphi}_{\mathrm{f,s}}=0 \tag{3.1.7}$$

$$\nabla^2\boldsymbol{\psi}+k_{\mathrm{t}}^2\boldsymbol{\psi}=0 \tag{3.1.8}$$

where $k_{\mathrm{f,s}}$ and k_{t} are the wavenumbers of body waves; $\varphi=\varphi_{\mathrm{f}}+\varphi_{\mathrm{s}}$, $\tilde{\chi}_{\mathrm{f,s}}=\xi_{\mathrm{f,s}}\tilde{\varphi}_{\mathrm{f,s}}$,$\tilde{\boldsymbol{\Theta}}=\xi_{\mathrm{t}}\tilde{\boldsymbol{\psi}}$,

$\xi_{\mathrm{f,s}}=\dfrac{(\lambda+\alpha^2M+2\mu)k_{\mathrm{f,s}}^2-\rho\omega^2}{\rho_{\mathrm{f}}\omega^2-\alpha Mk_{\mathrm{f,s}}^2}$, $\xi_{\mathrm{t}}=-\dfrac{\rho_{\mathrm{f}}\omega^2}{m\omega^2-\mathrm{i}b\omega}$.

The solutions for Eqs. (3.1.7) and (3.1.8) are

$$\begin{aligned}&\tilde{\varphi}_{\mathrm{f,s}}=\sum_{n=0}^{\infty}B_{1n,2n}K_n(\alpha_{1,2}r)\cos n\theta,\ \tilde{\psi}_r=\sum_{n=0}^{\infty}B_{rn}K_{n+1}(\beta r)\sin n\theta\\ &\tilde{\psi}_{\theta n}=-\sum_{n=0}^{\infty}B_{rn}K_{n+1}(\beta r)\cos n\theta,\ \tilde{\psi}_z=\sum_{n=0}^{\infty}B_{zn}K_n(\beta r)\sin n\theta\end{aligned} \tag{3.1.9}$$

where $\alpha_1^2=\xi^2-k_{\mathrm{f}}^2$, $\alpha_2^2=\xi^2-k_{\mathrm{s}}^2$ and $\beta^2=\xi^2-k_{\mathrm{t}}^2$; B_{1n}, B_{2n}, B_{rn}, and B_{zn} are the unknowns.

The displacements and stresses (including the pore pressure) are

$$\tilde{\boldsymbol{u}}=\begin{Bmatrix}\tilde{u}_r\\ \tilde{u}_\theta\\ \tilde{u}_z\end{Bmatrix}=\sum_{n=0}^{\infty}[\boldsymbol{S}]\cdot[\boldsymbol{U}]\cdot\boldsymbol{D}\text{ , }\tilde{w}_r=\sum_{n=0}^{\infty}[\boldsymbol{W}]\cdot\boldsymbol{D}\cdot\cos n\theta$$

$$\tilde{\boldsymbol{\sigma}}=\begin{Bmatrix}\tilde{\sigma}\\ \tilde{\sigma}_{\ \theta}\\ \tilde{\sigma}\\ \tilde{\sigma}_{\theta\theta}\\ \tilde{\sigma}_\theta\\ \tilde{\sigma}\end{Bmatrix}=\sum_{=0}^{\infty}\begin{bmatrix}\boldsymbol{S} & \boldsymbol{0}\\ \boldsymbol{0} & \boldsymbol{S}\end{bmatrix}\cdot[\boldsymbol{T}]\cdot\boldsymbol{D}\text{ , }\tilde{p}_{\mathrm{f}}=\sum_{n=0}^{\infty}[\boldsymbol{P}]\cdot\boldsymbol{D}\cdot\cos n\theta \tag{3.1.10}$$

$$[\boldsymbol{S}]=\begin{bmatrix}\cos n\theta & 0 & 0\\ 0 & \sin n\theta & 0\\ 0 & 0 & \cos n\theta\end{bmatrix}$$

where $\boldsymbol{D}=\{B_{1n},\ B_{2n},\ B_{rn},\ B_{zn}\}^{\mathrm{T}}$ is the unknown vector; matrices of $[\boldsymbol{U}]$ $[\boldsymbol{W}]$ $[\boldsymbol{T}]$ and $[\boldsymbol{P}]$ are given in the Appendix B. The fluid displacement components $\tilde{w}_\theta$ and $\tilde{w}_z$ are omitted for brevity.

In the transformed domain for a specified mode n, the displacements and stresses (including the pore pressure) are

$$\begin{Bmatrix}\tilde{U}_{rn}\\ \tilde{U}_{\theta n}\\ \tilde{U}_{zn}\end{Bmatrix}=[\boldsymbol{U}]\cdot\boldsymbol{D}\text{, }\tilde{W}_{rn}=[\boldsymbol{W}]\cdot\boldsymbol{D}\text{, }\begin{Bmatrix}\tilde{T}_{rrn}\\ \tilde{T}_{r\theta n}\\ \tilde{T}_{rzn}\end{Bmatrix}=[\boldsymbol{T}_r]\cdot\boldsymbol{D}\text{, }\tilde{P}_{\mathrm{f}n}=[\boldsymbol{P}]\cdot\boldsymbol{D} \tag{3.1.11}$$

where $[\boldsymbol{T}_r]$ is the first three rows of $[\boldsymbol{T}]$.

3.1.2 Solution for a Thin Shell Coupled with the Poroelastic Full-Space

The governing equations of a thin shell are

$$-v\frac{\partial u}{\partial z}-\frac{1}{a}\left(\frac{\partial \upsilon}{\partial\theta}+w\right)-\frac{h^2}{12}\left(a\frac{\partial^4 w}{\partial z^4}+\frac{2}{a}\frac{\partial^4 w}{\partial z^2\partial\theta^2}+\frac{1}{a^3}\frac{\partial^4 w}{\partial\theta^4}\right)$$

$$-\frac{h^2}{12}\left(-\frac{\partial^3 u}{\partial z^3}+\frac{1-v}{2a^2}\frac{\partial^3 u}{\partial z\partial\theta^2}-\frac{3-v}{2a}\frac{\partial^3\upsilon}{\partial z^2\partial\theta}+\frac{1}{a^3}w+\frac{2}{a^3}\frac{\partial^2 w}{\partial\theta^2}\right)+a\frac{1-v^2}{Eh}q_r-\rho_{\mathrm{t}}a\frac{1-v^2}{E}\frac{\partial^2 w}{\partial t^2}=0 \tag{3.1.12}$$

$$\frac{1+v}{2}\frac{\partial^2 u}{\partial z\partial\theta}+a\frac{1-v}{2}\frac{\partial^2\upsilon}{\partial z^2}+\frac{1}{a}\left(\frac{\partial^2\upsilon}{\partial\theta^2}+\frac{\partial w}{\partial\theta}\right)+\frac{h^2}{12}\left[\frac{3(1-v)}{2a}\frac{\partial^2\upsilon}{\partial z^2}-\frac{3-v}{2a}\frac{\partial^3 w}{\partial z^2\partial\theta}\right]$$

$$+a\frac{1-v^2}{Eh}q_\theta-\rho_{\mathrm{t}}a\frac{1-v^2}{E}\frac{\partial^2\upsilon}{\partial t^2}=0 \tag{3.1.13}$$

$$a\frac{\partial^2 u}{\partial z^2}+\frac{1-v}{2a}\frac{\partial^2 u}{\partial \theta^2}+\frac{1+v}{2}\frac{\partial^2 \upsilon}{\partial z\partial \theta}+v\frac{\partial w}{\partial z}+\frac{h^2}{12}\left(\frac{1-v}{2a^3}\frac{\partial^2 u}{\partial \theta^2}-\frac{\partial^3 w}{\partial z^3}+\frac{1-v}{2a^2}\frac{\partial^3 w}{\partial z\partial \theta^2}\right)$$

$$+a\frac{1-v^2}{Eh}q_z-\rho_t a\frac{1-v^2}{E}\frac{\partial^2 u}{\partial t^2}=0 \tag{3.1.14}$$

Here, u, υ and w are the displacement components. The tunnel has Young's modulus E, Poisson's ratio v, the density ρ_t and the damping ratio β_t. q_r, q_θ, q_z are the net applied loads.

The net applied loads can be expressed as a combination of trigonometric terms

$$q_r(z,t)=\sum_{n=0}^{\infty}Q_{rn}\cos n\theta,\quad q_\theta(z,t)=\sum_{n=0}^{\infty}Q_{\theta n}\sin n\theta,\quad q_z(z,t)=\sum_{n=0}^{\infty}Q_{zn}\cos n\theta \tag{3.1.15}$$

$$w(z,t)=\sum_{n=0}^{\infty}W_n\cos n\theta,\quad \upsilon(z,t)=\sum_{n=0}^{\infty}V_n\sin n\theta,\quad u(z,t)=\sum_{n=0}^{\infty}U_n\cos n\theta \tag{3.1.16}$$

where Q_{rn}, $Q_{\theta n}$ and Q_{zn} are the load amplitudes for a mode n; W_n, V_n and U_n are the displacement amplitudes for a mode n.

In combination with Fourier transform, the substitution of Eqs. (3.1.15) and (3.1.16) into Eqs. (3.1.12)–(3.1.14) yields

$$[\boldsymbol{A}]\begin{Bmatrix}\widetilde{W}_n\\ \widetilde{V}_n\\ \widetilde{U}_n\end{Bmatrix}=\frac{-a(1-v^2)}{Eh}\begin{Bmatrix}\widetilde{Q}_{rn}\\ \widetilde{Q}_{\theta n}\\ \widetilde{Q}_{zn}\end{Bmatrix} \tag{3.1.17}$$

where $[\boldsymbol{A}]$ is given in the Appendix B.

If a unit point load is applied at the invert

$$p_r=\frac{\delta(t)\delta(\theta)\delta(z)}{a},\quad p_\theta=p_z=0 \tag{3.1.18}$$

It can be extended into Fourier series and expressed in the transformed domain

$$\widetilde{p}_r=\frac{1}{(1+\delta_{n0})a\pi}\sum_{n=0}^{\infty}\cos n\theta,\quad \widetilde{p}_\theta=\widetilde{p}_z=0 \tag{3.1.19}$$

where δ_{n0}=1 for n=0 and δ_{n0}=0 for n>0.

Removing the cosine functions, the stresses at the inner surface of the tunnel for a mode n are

$$\widetilde{P}_{rn}=\frac{1}{(1+\delta_{n0})a\pi},\quad \widetilde{P}_{\theta n}=\widetilde{P}_{zn}=0 \tag{3.1.20}$$

The applied load in Eq. (3.1.17) is the stress difference between the inner and outer surfaces of the tunnel

$$\frac{-Eh}{a(1-v^2)}[\boldsymbol{A}]\begin{Bmatrix}\widetilde{W}_n\\ \widetilde{V}_n\\ \widetilde{U}_n\end{Bmatrix}=\begin{Bmatrix}\widetilde{Q}_{rn}\\ \widetilde{Q}_{\theta n}\\ \widetilde{Q}_{zn}\end{Bmatrix}=\begin{Bmatrix}\widetilde{P}_{rn}\\ \widetilde{P}_{\theta n}\\ \widetilde{P}_{zn}\end{Bmatrix}-\begin{Bmatrix}\widetilde{T}_{rn}\\ \widetilde{T}_{\theta n}\\ \widetilde{T}_{zn}\end{Bmatrix}_{\text{outside}} \tag{3.1.21}$$

where $\left\{\widetilde{T}_{rn},,\widetilde{T}_{\theta n},\widetilde{T}_{zn}\right\}^{\mathrm{T}}_{\text{outside}}$ is the stress vector at the tunnel-soil interface.

The displacement continuity condition at the tunnel-soil interface is

$$\begin{Bmatrix} \widetilde{W}_n \\ \widetilde{V}_n \\ \widetilde{U}_n \end{Bmatrix} = \begin{Bmatrix} \widetilde{U}_{rn} \\ \widetilde{U}_{\theta n} \\ \widetilde{U}_{zn} \end{Bmatrix}_{r=a} = [\boldsymbol{U}]_{r=a} \cdot \boldsymbol{D} \tag{3.1.22}$$

The stress continuity condition at the tunnel-soil interface is

$$\begin{Bmatrix} \widetilde{T}_{rn} \\ \widetilde{T}_{\theta n} \\ \widetilde{T}_{zn} \end{Bmatrix}_{\text{outside}} = \begin{Bmatrix} \widetilde{T}_{rrn} \\ \widetilde{T}_{r\theta n} \\ \widetilde{T}_{rzn} \end{Bmatrix}_{r=a} = [\boldsymbol{T}_r]_{r=a} \cdot \boldsymbol{D} \tag{3.1.23}$$

An impervious boundary condition at the tunnel-soil interface is

$$\widetilde{W}_{rn}\Big|_{r=a} = [\boldsymbol{W}]_{r=a} \cdot \boldsymbol{D} = 0 \tag{3.1.24}$$

Solving Eqs. (3.1.21)-(3.1.24), the unknown vector is

$$\boldsymbol{D} = \begin{bmatrix} \dfrac{-Eh}{a(1-\nu^2)}[\boldsymbol{A}]\cdot[\boldsymbol{U}]_{r=a} - [\boldsymbol{T}_r]_{r=a} \\ [\boldsymbol{W}]_{r=a} \end{bmatrix}^{-1} \begin{Bmatrix} \widetilde{P}_{rn} \\ \widetilde{P}_{\theta n} \\ \widetilde{P}_{zn} \\ 0 \end{Bmatrix} \tag{3.1.25}$$

The displacement components and the pore pressure for a mode n are

$$\begin{Bmatrix} \widetilde{U}_{rn} \\ \widetilde{U}_{\theta n} \\ \widetilde{U}_{zn} \\ \widetilde{P}_{\text{f}n} \end{Bmatrix}_{r=R} = \begin{bmatrix} [\boldsymbol{U}]_{r=R} \\ [\boldsymbol{P}]_{r=R} \end{bmatrix} \cdot \boldsymbol{D} \tag{3.1.26}$$

The transfer functions of displacement components and the pore pressure are obtained from a sufficient number of the modal components

$$\begin{Bmatrix} \widetilde{u}_r \\ \widetilde{u}_\theta \\ \widetilde{u}_z \\ \widetilde{p}_\text{f} \end{Bmatrix} = \sum_{n=0}^{\infty} \begin{Bmatrix} \widetilde{U}_{rn}(r,\xi,\omega)\cos n\theta \\ \widetilde{U}_{\theta n}(r,\xi,\omega)\sin n\theta \\ \widetilde{U}_{zn}(r,\xi,\omega)\cos n\theta \\ \widetilde{P}_{\text{f}n}(r,\xi,\omega)\cos n\theta \end{Bmatrix} \tag{3.1.27}$$

3.1.3 Ground Vibrations Induced by the Metro Lines

With an increasing traffic volume in recent years, the subway becomes an effective measure to relieve the traffic congestion; however, underground railways exerts significant negative influences on the nearby buildings and residents due to the vibration and noise pollution. It causes anxiety, deprives people's sleep and interferes with normal day-to-day life, which has been recognized as an environmental issue. Thus it is important to calculate the ground-borne vibrations generated by the underground moving trains accurately.

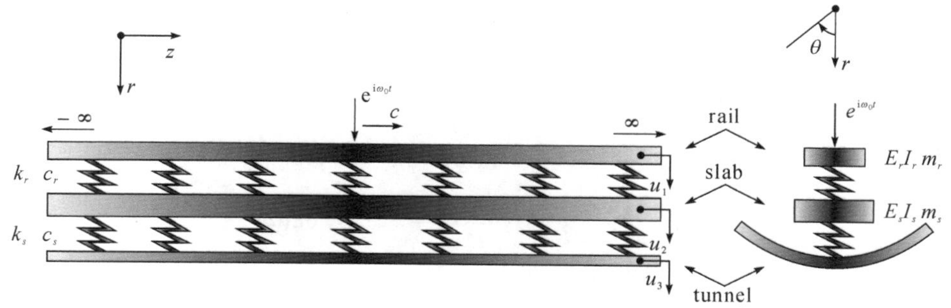

Fig. 3.1.2 A floating-slab track model

Referring to the study by Hussien et al.[16], the two rails of the track and the slab are modelled as Euler-Bernoulli beams resting on the tunnel invert as shown in Fig. 3.1.2.

The governing equations of the track are

$$E_r I_r \frac{\partial^4 u_1}{\partial z^4} + m_r \frac{\partial^2 u_1}{\partial t^2} + k_r(u_1 - u_2) + c_r(\frac{\partial u_1}{\partial t} - \frac{\partial u_2}{\partial t}) = e^{i\omega_0 t}\delta(z - ct) \tag{3.1.28}$$

$$E_s I_s \frac{\partial^4 u_2}{\partial z^4} + m_s \frac{\partial^2 u_2}{\partial t^2} - k_r(u_1 - u_2) - c_r(\frac{\partial u_1}{\partial t} - \frac{\partial u_2}{\partial t}) + F_R = 0 \tag{3.1.29}$$

$$F_R = k_s(u_2 - u_3) + c_s(\frac{\partial u_2}{\partial t} - \frac{\partial u_3}{\partial t}) \tag{3.1.30}$$

where u_1, u_2 and u_3 are the displacements of the rail, the slab and the tunnel invert; the subscripts r and s denote the rail and the slab, respectively; EI is the bending stiffness; m is the mass; k and c are the stiffness and the damping factor, respectively; F_R is compressive force applied at the tunnel invert; $f_0=\omega_0/2\pi$ and c are the load frequency and speed, respectively.

In Sections 3.1.1 and 3.1.2, the transfer functions of displacement components and the pore pressure are obtained, which could be used to couple with the track structures and calculate the vibrations from underground railways. The vertical displacement at the tunnel invert, u_3 as shown in Fig. 3.1.2, is

$$\widetilde{u}_3 = H\widetilde{F}_R \tag{3.1.31}$$

where H is the transfer function of the vertical displacement at the tunnel invert.

Solving Eqs. (3.1.28)–(3.1.31), the following equations can be obtained

$$\widetilde{u}_1 = \frac{D_3(1 + D_4 H) + D_4}{D_5}\widetilde{F} \tag{3.1.32}$$

$$\widetilde{u}_2 = \frac{D_2(1 + D_4 H)}{D_5}\widetilde{F} \tag{3.1.33}$$

$$\widetilde{u}_3 = \frac{D_2 D_4 H}{D_5}\widetilde{F} \tag{3.1.34}$$

$$\widetilde{F}_R = \frac{D_2 D_4}{D_5}\widetilde{F} \tag{3.1.35}$$

where $D_1 = E_r I_r \xi^4 - m_r \omega^2 + k_r + c_r \mathrm{i}\omega$, $D_2 = k_r + c_r \mathrm{i}\omega$, $D_3 = E_s I_s \xi^4 - m_s \omega^2 + k_r + c_r \mathrm{i}\omega$, $D_4 = k_s + c_s \mathrm{i}\omega$, $D_5 = (D_1 D_3 - D_2^2)(1 + D_4 H) + D_1 D_4$, $\widetilde{F} = 2\pi\delta(\omega + \xi c - \omega_0)$.

The results in the physical domain is

$$u_1(z,t) = \frac{1}{4\pi^2}\int_{-\infty}^{\infty}\int_{-\infty}^{\infty} \frac{D_3(1 + D_4 H) + D_4}{D_5} F \mathrm{e}^{\mathrm{i}(\xi z + \omega t)} \mathrm{d}\xi \mathrm{d}\omega \tag{3.1.36}$$

$$u_2(z,t) = \frac{1}{4\pi^2}\int_{-\infty}^{\infty}\int_{-\infty}^{\infty} \frac{D_2(1 + D_4 H)}{D_5} F \mathrm{e}^{\mathrm{i}(\xi z + \omega t)} \mathrm{d}\xi \mathrm{d}\omega \tag{3.1.37}$$

$$u_3(z,t) = \frac{1}{4\pi^2}\int_{-\infty}^{\infty}\int_{-\infty}^{\infty} \frac{D_2 D_4 H}{D_5} F \mathrm{e}^{\mathrm{i}(\xi z + \omega t)} \mathrm{d}\xi \mathrm{d}\omega \tag{3.1.38}$$

Replacing H in Eq. (3.1.38) with H_r, H_θ, H_z and H_p, the displacement and the pore pressure at arbitrary positions are

$$\begin{Bmatrix} u_r \\ u_\theta \\ u_z \\ p_\mathrm{f} \end{Bmatrix} = \frac{1}{4\pi^2}\int_{-\infty}^{\infty}\int_{-\infty}^{\infty} \begin{Bmatrix} H_r \\ H_\theta \\ H_z \\ H_\mathrm{p} \end{Bmatrix} \frac{D_2 D_4}{D_5} \widetilde{F} \mathrm{e}^{\mathrm{i}(\omega t + \xi z)} \mathrm{d}\xi \mathrm{d}\omega \tag{3.1.39}$$

where H_r, H_θ, H_z and H_p are the transfer functions calculated by Eq. (3.1.27) at arbitrary positions.

Validation of the Tunnel-Soil Model

Results calculated by the proposed model are compared with those obtained by Lu[17] (unlined tunnel) and by Hasheminejad[18] (lined tunnel) in Fig. 3.1.3. In the works of Lu and Hasheminejad, a moving axisymmetric ring load is applied at the tunnel invert, which is different from the case in the present work. But the dynamic responses to a ring load can be approximated by the superimposition of the responses to discrete point loads. Fig. 3.1.3A compares the solid radial displacement at r=1.5a calculated by the present model with that obtained by Lu with the same material parameters. It can be seen that a good correspondence is observed at c=0.1 V_s and c=0.9 V_s. Fig. 3.1.3B shows the solid radial displacement at r=4.5 m for two load velocities c=10 m/s and 100 m/s. The material parameters of the saturated soil are set to the values of stiff sandstone used by Hasheminejad [18]. It can be observed from Fig. 3.1.3B that two results are in good agreement again.

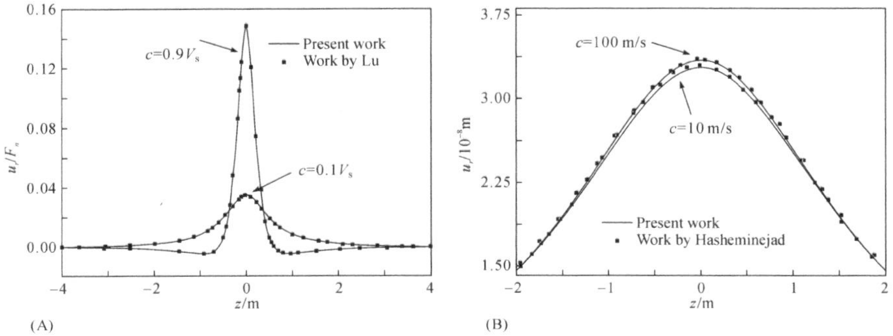

Fig. 3.1.3 Comparison of the present work with existing works. (A) Lu[18]; (B) Hasheminejad[19]

Numerical Results and Discussion

The parameters for the track are listed in Table 3.1.1, which are the same as those used in Hussein et al.[16] The material and geometrical parameters for the saturated soil and the tunnel are presented in Table 3.1.2, which are selected from the work Yuan et al.[19]

Table 3.1.1 Parameters of the floating slab tracks

Parameter	Value
Bending stiffness of the two rails, E_rI_r	1×10^7 Pa·m^4
Bending stiffness of the slab, E_sI_s	1.43×10^9 Pa·m^4
Mass per unit length of the two rails, m_r	100 kg/m
Mass per unit length of the slab, m_s	3500 kg/m
Stiffness of the rail pads, k_r	4×10^7 N/m^2
Stiffness of the slab bearing, k_s	5×10^7 N/m^2
Damping factors of the rail pads, c_r	6.3×10^3 N·s/m^2
Damping factors of the slab bearing, c_s	4.18×10^4 N·s/m^2

Table 3.1.2 Parameters of the saturated soil and tunnel

Parameter	Value
Lame constant, μ	2×10^7 N/m^2
Lame constant, λ	4.67×10^7 N/m^2
Solid density, ρ_s	1816 kg/m^3
Water density, ρ_f	1000 kg/m^3
Porosity, n	0.4
Ratio between the fluid viscosity and intrinsic permeability, b	$1\times10^{6\text{-}9}$ kg/(m^{3}s)
Hysteretic material damping in the soil, β	0.1
Parameter for compressibility of the soil particle, α	1.0
Parameter for compressibility of the fluid, M	6.125×10^9 N/m^2
Young's modulus of the shell, E	5×10^{10} N/m^2
Poisson's ratio of the shell, v	0.3
Density of the shell, ρ_t	2500 kg/m^3
Hysteretic material damping in the shell, β_t	0.02
Radius of the shell, a	3 m
Thickness of the shell, h	0.25 m

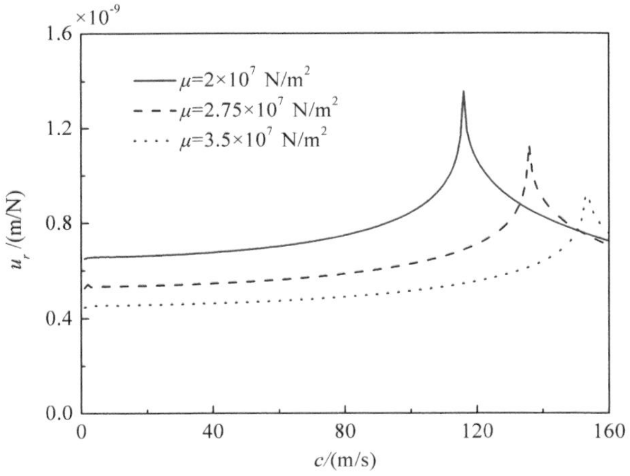

Fig. 3.1.4 Effects of the solid skeleton modulus on the soil displacement to a unit constant load moving on the tunnel invert

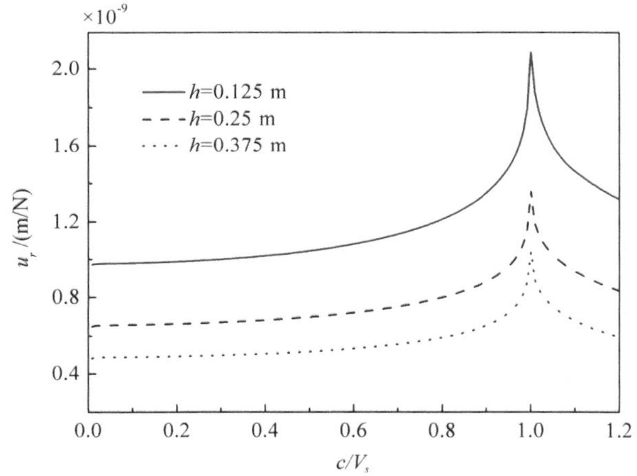

Fig. 3.1.5 Effects of the tunnel thickness on the soil displacement to a unit constant load moving on the tunnel invert

Fig. 3.1.4 presents dynamic response of the saturated soil due to a unit moving constant point load without floating slab tracks. The observation point is chosen at r=6 m, θ=0 and z=0 in the free field. Fig. 3.1.4 plots the radial solid displacement amplitude u_r against the load velocity c for three different values of solid shear modulus μ. As expected, with decreasing shear modulus μ, the displacement response generally increases because of the decreased stiffness of the tunnel-soil system. It is interesting to note that each curve in Fig. 3.1.4 has the similar varying tendency, i.e., the displacement increases with the load velocity and reaches a maximum, then decreases gradually as the velocity increases further. Here we define critical velocity as the velocity at which the displacement under the load reaches a maximum. It is found the critical velocity is equal to the undrained shear wave speed of the saturated soil ($V_s=\sqrt{\mu/\rho}$=116 m/s for $\mu=2\times10^7$ N/m^2, 136 m/s for $\mu=2.75\times10^7$ N/m^2 and 153 m/s for $\mu=3.5\times10^7$ N/m^2). From Fig. 3.1.5, it is seen that the stiffness of the tunnel-soil system increases with the increasing tunnel thickness h, but the critical velocity of the tunnel-soil system does not depend on the tunnel thickness h.

Floating slab track is widely used as effective means for vibration isolation in underground tunnels. In order to investigate the effect of the track system on the ground deformation, the soil displacement at r=6 m, θ=0 and z=0 for the cases with continuous floating slab track and without the track is presented in Fig. 3.1.6. It is observed in Fig. 3.1.6 the maximum displacement due to a constant moving load for the two cases appears at c/V_s=1.0 (critical velocity of the tunnel-soil system, $\mu=2\times10^7$ N/m^2). The installation of floating slab tracks does not alter the critical velocity of the system. Generally, the soil displacement for the case with track is smaller than that without the track due to the fact that mounting a floating slab track on the tunnel invert allows the energy to propagate along the track before being transmitted into the tunnel and the soil.

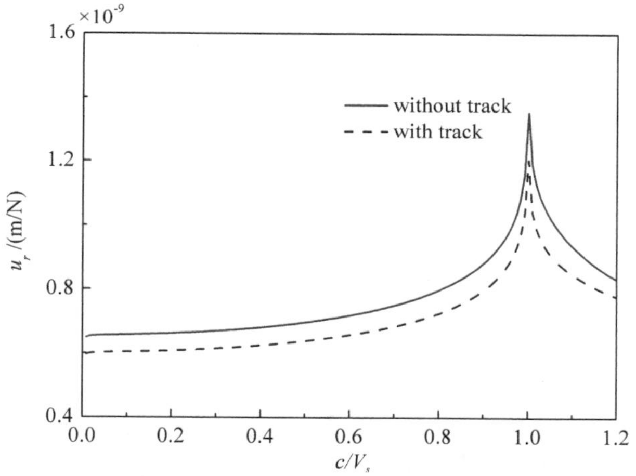

Fig. 3.1.6 Comparison between the cases with floating slab track and without floating slab track to a moving constant load

Fig. 3.1.7 compares the maximum displacement amplitude in the free field due to a unit harmonic moving load (f_0=1–200 Hz, c=40 km/h) for the two cases. The response magnitude is presented in decibels ([dB]= 20log(P1/P2) in which P1 is the computed response amplitude, and P2 is the reference value). The displacement curve for the case without track shows an undulating characteristic due to the dynamic tunnel-soil interaction. For the case with track, a resonance peak can be observed at f_0=19.02 Hz, which is the natural frequency of the slab calculated from a single degree system ($(1/(2\pi))\sqrt{k_s/m_s}$=19.02 Hz). The isolation effect is not achieved at load frequencies around the natural frequency of the slab but it is observed when f_0>25 Hz, as shown in Fig. 3.1.7.

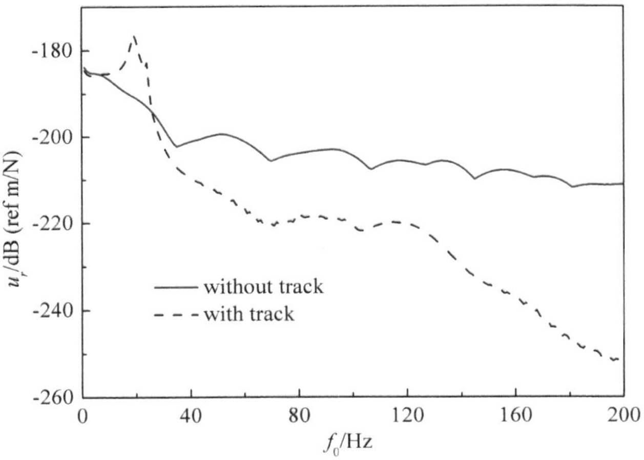

Fig. 3.1.7 Comparison between the cases with and without floating slab track (c=40 km/h)

In this subsection, the dynamic responses of floating slab tracks and underground railway tunnel embedded in a three-dimensional poroelastic full-space induced by a unit moving point load are investigated theoretically. The following conclusions can be drawn: the free-field responses generated by a constant moving load present resonance characteristics and the installation of floating slab tracks does not alter the critical velocity of the system; the natural frequency of the floating slab has significant influence on the soil displacement in the free field and if the load frequency is higher than the natural frequency of the slab, obvious isolation effect is achieved.

3.2 Dynamic Response of a Tunnel Embedded in a Saturated Poroelastic Half-Space

The wave field in the half-space surrounding the tunnel is composed of down-going waves and outgoing waves. In Fig. 3.2.1, the tunnel is modelled as an elastic cylinder with inner radius a, outer radius b and a depth d.

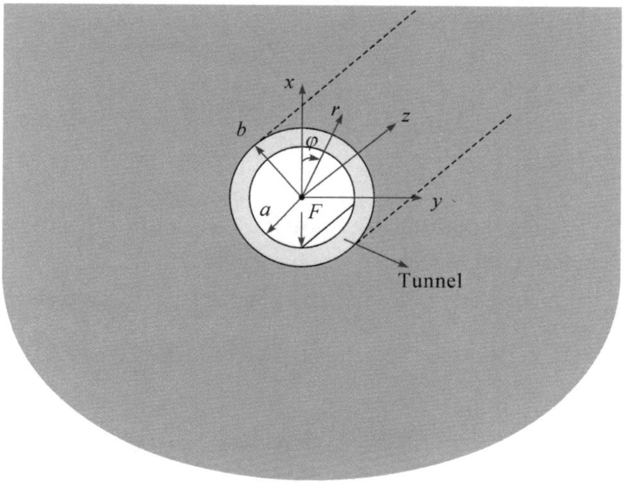

Fig. 3.2.1 A circular tunnel embedded in the saturated poroelastic half-space

3.2.1 Solution for a Cavity in the Saturated Poroelastic Half-space

In Cartesian coordinates, the down-going and up-going waves constitute the solutions of Eqs. (1.1.24)–(1.1.25). The down-going waves are

$$\phi_1^- = \nabla\times(\mathbf{e}_z e^{i(qz-h_s x)}\sin py) = (p\cos py, ih_s \sin py, 0)e^{i(qz-h_s x)} \tag{3.2.1}$$

$$\phi_2^- = \nabla\times\nabla\times(\mathbf{e}_z e^{i(qz-h_s x)}\cos py) = (qh_s \cos py, iqp\sin py, (k_s^2 - q^2)\cos py)e^{i(qz-h_s x)} \tag{3.2.2}$$

$$\phi_3^- = \nabla(e^{i(qz-h_{p1}x)}\cos py) = (-ih_{p1}\cos py, -p\sin py, iq\cos py)e^{i(qz-h_{p1}x)} \tag{3.2.3}$$

$$\phi_4^- = \nabla(e^{i(qz-h_{p2}x)}\cos py) = (ih_{p2}\cos py, p\sin py, iq\cos py)e^{i(qz-h_{p2}x)} \tag{3.2.4}$$

where ϕ_j^- are the wave potentials for the down-going waves; k_s, k_{p1}, k_{p2} are the wavenumbers for S, P1 and P2 waves, respectively; h_s, h_{p1}, h_{p2}, p and q are the wavenumber components in x, y and z directions, respectively. A sign change in the vertical wavenumber yields the wave potentials for the up-going waves ϕ_j^+.

The stress vector for down-going waves is

$$\boldsymbol{t}^{(\mathbf{e}_x)}(\phi_1^-) = \mu(-2iph_s\cos py, (h_s^2 - p^2)\sin py, iqp\cos py)e^{i(qz-h_s x)} \tag{3.2.5}$$

$$\boldsymbol{t}^{(\mathbf{e}_x)}(\phi_2^-) = \mu(-2iqh_s^2\cos py, -2qph_s\sin py, -ih_s(k_s^2 - 2q^2)\cos py)e^{i(qz-h_s x)} \tag{3.2.6}$$

$$\begin{aligned}\boldsymbol{t}^{(\mathbf{e}_x)}(\phi_3^-) = \mu(&(2k_{p1}^2 - k_{p1}^2(\frac{\lambda}{\mu}+2) - 2h_{p1}^2 - \alpha(\alpha+\xi_{p1})Mk_{p1}^2)\cos py, 2iph_{p1}\sin py,\\ &2qh_{p1}\cos py)e^{i(qz-h_{p1}x)}\end{aligned} \tag{3.2.7}$$

$$\begin{aligned}\boldsymbol{t}^{(\mathbf{e}_x)}(\phi_4^-) = \mu(&(2k_{p2}^2 - k_{p2}^2(\frac{\lambda}{\mu}+2) - 2h_{p2}^2 - \alpha(\alpha+\xi_{p2})Mk_{p2}^2)\cos py,\\ &2iph_{p2}\sin py, 2qh_{p2}\cos py)e^{i(qz-h_{p2}x)}\end{aligned} \tag{3.2.8}$$

A sign change in the vertical wavenumber yields the stress vector for up-going waves $\boldsymbol{t}^{(\mathbf{e}_x)}(\phi_j^+)$.

The pore pressure for down-going waves is

$$p_{\mathrm{f}}(\phi_3^-)=\alpha(\alpha+\xi_{\mathrm{p1}})Mk_{\mathrm{p1}}^2\cos py\mathrm{e}^{\mathrm{i}(qz-h_{\mathrm{p1}}x)} \tag{3.2.9}$$

$$p_{\mathrm{f}}(\phi_4^-)=\alpha(\alpha+\xi_{\mathrm{p2}})Mk_{\mathrm{p2}}^2\cos py\mathrm{e}^{\mathrm{i}(qz-h_{\mathrm{p2}}x)} \tag{3.2.10}$$

where $\xi_{\mathrm{s}}=-\dfrac{\rho_{\mathrm{f}}\omega^2}{m\omega^2+\mathrm{i}b\omega}$, $\xi_{\mathrm{p1,p2}}=\dfrac{(\lambda+\alpha^2M+2\mu)k_{\mathrm{p1,p2}}^2-\rho\omega^2}{\rho_{\mathrm{f}}\omega^2-\alpha Mk_{\mathrm{p1,p2}}^2}$.

A sign change in the vertical wavenumber yields the pore pressure for up-going waves $p_{\mathrm{f}}(\phi_j^+)$.

In cylindrical coordinates, the outgoing waves are

$$\chi_{1m}^+=\nabla\times(\mathbf{e}_zH_m^{(1)}(g_{\mathrm{s}}r)\sin m\varphi\mathrm{e}^{\mathrm{i}qz})=(\frac{m}{r}H_m^{(1)}(g_{\mathrm{s}}r)\cos m\varphi,-g_{\mathrm{s}}H_m^{(1)\prime}(g_{\mathrm{s}}r)\sin m\varphi,0)\mathrm{e}^{\mathrm{i}qz} \tag{3.2.11}$$

$$\chi_{2m}^+=\nabla\times\nabla\times(\mathbf{e}_zH_m^{(1)}(g_{\mathrm{s}}r)\cos m\varphi\mathrm{e}^{\mathrm{i}qz})=(\mathrm{i}qg_{\mathrm{s}}H_m^{(1)\prime}(g_{\mathrm{s}}r)\cos m\varphi,\frac{-\mathrm{i}mq}{r}H_m^{(1)}(g_{\mathrm{s}}r)\sin m\varphi,$$
$$g_{\mathrm{s}}^2H_m^{(1)}(g_{\mathrm{s}}r)\cos m\varphi)\mathrm{e}^{\mathrm{i}qz} \tag{3.2.12}$$

$$\chi_{3m}^+=\nabla(H_m^{(1)}(g_{\mathrm{p1}}r)\cos m\varphi\mathrm{e}^{\mathrm{i}qz})=(g_{\mathrm{p1}}H_m^{(1)\prime}(g_{\mathrm{p1}}r)\cos m\varphi,-\frac{m}{r}H_m^{(1)}(g_{\mathrm{p1}}r)\sin m\varphi,$$
$$\mathrm{i}qH_m^{(1)}(g_{\mathrm{p1}}r)\cos m\varphi)\mathrm{e}^{\mathrm{i}qz} \tag{3.2.13}$$

$$\chi_{4m}^+=\nabla(H_m^{(1)}(g_{\mathrm{p2}}r)\cos m\varphi\mathrm{e}^{\mathrm{i}qz})=(g_{\mathrm{p2}}H_m^{(1)\prime}(g_{\mathrm{p2}}r)\cos m\varphi,-\frac{m}{r}H_m^{(1)}(g_{\mathrm{p2}}r)\sin m\varphi,$$
$$\mathrm{i}qH_m^{(1)}(g_{\mathrm{p2}}r)\cos m\varphi)\mathrm{e}^{\mathrm{i}qz} \tag{3.2.14}$$

where χ_{jm}^+ are the wave potentials for outgoing waves; $g_{\mathrm{s,\,p1,\,p2}}$ are the wavenumbers in the r direction. A replacement of $H_m^{(1)}$ with J_m in Eqs. (3.2.11)–(3.2.14) gives the wave potentials for regular waves.

The stress vector for outgoing waves is

$$\boldsymbol{t}^{(\mathbf{e}_r)}(\chi_{1m}^+)=\mu((\frac{2mg_{\mathrm{s}}}{r}H_m^{(1)\prime}(g_{\mathrm{s}}r)-\frac{2m}{r^2}H_m^{(1)}(g_{\mathrm{s}}r))\cos m\varphi,-g_{\mathrm{s}}^2(2H_m^{(1)\prime\prime}(g_{\mathrm{s}}r)+H_m^{(1)}(g_{\mathrm{s}}r))\sin m\varphi,$$
$$\frac{\mathrm{i}mq}{r}H_m^{(1)}(g_{\mathrm{s}}r)\cos m\varphi)\mathrm{e}^{\mathrm{i}qz} \tag{3.2.15}$$

$$\boldsymbol{t}^{(\mathbf{e}_r)}(\chi_{2m}^+)=\mu(2\mathrm{i}qg_{\mathrm{s}}^2H_m^{(1)\prime\prime}(g_{\mathrm{s}}r)\cos m\varphi,2\mathrm{i}mq(\frac{1}{r^2}H_m^{(1)}(g_{\mathrm{s}}r)-\frac{g_{\mathrm{s}}}{r}H_m^{(1)\prime}(g_{\mathrm{s}}r))\sin m\varphi,$$
$$g_{\mathrm{s}}(k_{\mathrm{s}}^2-2q^2)H_m^{(1)\prime}(g_{\mathrm{s}}r)\cos m\varphi)\mathrm{e}^{\mathrm{i}qz} \tag{3.2.16}$$

$$\boldsymbol{t}^{(\mathbf{e}_r)}(\chi_{3m}^+)=\mu(((2k_{\mathrm{p1}}^2-k_{\mathrm{p1}}^2(\frac{\lambda}{\mu}+2))H_m^{(1)}(g_{\mathrm{p1}}r)+2g_{\mathrm{p1}}^2H_m^{(1)\prime\prime}(g_{\mathrm{p1}}r))\cos m\varphi,$$
$$2m(\frac{1}{r^2}H_m^{(1)}(g_{\mathrm{p1}}r)-\frac{g_{\mathrm{p1}}}{r}H_m^{(1)\prime}(g_{\mathrm{p1}}r))\sin m\varphi,2\mathrm{i}qg_{\mathrm{p1}}H_m^{(1)\prime}(g_{\mathrm{p1}}r)\cos m\varphi)\mathrm{e}^{\mathrm{i}qz} \tag{3.2.17}$$

$$\boldsymbol{t}^{(\mathbf{e}_r)}(\chi_{4m}^+)=\mu(((2k_{\mathrm{p2}}^2-k_{\mathrm{p2}}^2(\frac{\lambda}{\mu}+2))H_m^{(1)}(g_{\mathrm{p2}}r)+2g_{\mathrm{p2}}^2H_m^{(1)\prime\prime}(g_{\mathrm{p2}}r))\cos m\varphi,$$
$$2m(\frac{1}{r^2}H_m^{(1)}(g_{\mathrm{p2}}r)-\frac{g_{\mathrm{p2}}}{r}H_m^{(1)\prime}(g_{\mathrm{p2}}r))\sin m\varphi,2\mathrm{i}qg_{\mathrm{p2}}H_m^{(1)\prime}(g_{\mathrm{p2}}r)\cos m\varphi)\mathrm{e}^{\mathrm{i}qz} \tag{3.2.18}$$

A replacement of $H_m^{(1)}$ with J_m in Eqs. (3.2.15)–(3.2.18) gives the stress vector for regular waves.

The pore pressure for outgoing waves is

$$p_{\mathrm{f}}(\chi_{3m}^+)=\alpha(\alpha+\xi_{\mathrm{p1}})Mk_{\mathrm{p1}}^2H_m^{(1)}(g_{\mathrm{p1}}r)\cos m\varphi\mathrm{e}^{\mathrm{i}qz} \tag{3.2.19}$$

$$p_{\mathrm{f}}(\chi_{4m}^+)=\alpha(\alpha+\xi_{\mathrm{p2}})Mk_{\mathrm{p2}}^2H_m^{(1)}(g_{\mathrm{p2}}r)\cos m\varphi\mathrm{e}^{\mathrm{i}qz} \tag{3.2.20}$$

A replacement of $H_m^{(1)}$ with J_m in Eqs. (3.2.19)–(3.2.20) gives the pore pressure for regular waves $p_{\mathrm{f}}(\chi_{jm}^0)$.

The wave field in the half-space surrounding the tunnel is

$$\boldsymbol{u}=\int_{-\infty}^{\infty}\mathrm{d}q\int_{0}^{\infty}\mathrm{d}p\sum_{j=1}^{4}A_j(q,p)\phi_j^{-}(q,p,\boldsymbol{x})+\int_{-\infty}^{\infty}\mathrm{d}q\sum_{j=1}^{4}\sum_{m=0}^{\infty}B_{jm}(q)\boldsymbol{\chi}_{jm}^{+}(q,\boldsymbol{r})$$
$$\boldsymbol{w}=\int_{-\infty}^{\infty}\mathrm{d}q\int_{0}^{\infty}\mathrm{d}p\sum_{j=1}^{4}\xi_j A_j(q,p)\phi_j^{-}(q,p,\boldsymbol{x})+\int_{-\infty}^{\infty}\mathrm{d}q\sum_{j=1}^{4}\sum_{m=0}^{\infty}\xi_j B_{jm}(q)\boldsymbol{\chi}_{jm}^{+}(q,\boldsymbol{r}) \tag{3.2.21}$$

$$p_{\mathrm{f}}(\boldsymbol{u},\boldsymbol{w})=\int_{-\infty}^{\infty}\mathrm{d}q\int_{0}^{\infty}\mathrm{d}p\sum_{j=3}^{4}A_j(q,p)p_{\mathrm{f}}(\phi_j^{-}(q,p,\boldsymbol{x}))+\int_{-\infty}^{\infty}\mathrm{d}q\sum_{j=3}^{4}\sum_{m=0}^{\infty}B_{jm}(q)p_{\mathrm{f}}(\boldsymbol{\chi}_{jm}^{+}(q,\boldsymbol{r})) \tag{3.2.22}$$

where A_j and B_{jm} are the unknowns.

The Hankel functions can be expressed as

$$H_m^{(1)}(g_{\mathrm{s}}r)\sin m\varphi=\frac{2\mathrm{i}^{-m}}{\pi}\int_0^{\infty}\mathrm{e}^{\mathrm{i}h_{\mathrm{s}}x}\sin py\sin m\alpha_{\mathrm{s}}\frac{1}{h_{\mathrm{s}}}\mathrm{d}p \tag{3.2.23}$$

$$H_m^{(1)}(g_{\mathrm{s,p1,p2}}r)\cos m\varphi=\frac{2\mathrm{i}^{-m}}{\pi}\int_0^{\infty}\mathrm{e}^{\mathrm{i}h_{\mathrm{s,p1,p2}}x}\cos py\cos m\alpha_{\mathrm{s,p1,p2}}\frac{1}{h_{\mathrm{s,p1,p2}}}\mathrm{d}p \tag{3.2.24}$$

From Eqs. (3.2.23)–(3.2.24), outgoing waves become

$$\boldsymbol{\chi}_{jm}^{+}(q,\boldsymbol{r})=\frac{2}{\pi}\int_0^{\infty}\phi_j^{+}(q,p,\boldsymbol{x})M_{jm}(p)\frac{1}{h_j}\mathrm{d}p \tag{3.2.25}$$

where $M_{jm}(p)=\mathrm{i}^{-m}\begin{cases}\sin m\alpha_j,\ j=1\\ \cos m\alpha_j,\ j=2,3,4\end{cases}$, $h_j=\begin{cases}h_{\mathrm{s}},\ j=1,2\\ h_{\mathrm{p1}},\ j=3\\ h_{\mathrm{p2}},\ j=4\end{cases}$, $\alpha_j=\begin{cases}\alpha_{\mathrm{s}},\ j=1,2\\ \alpha_{\mathrm{p1}},\ j=3\\ \alpha_{\mathrm{p2}},\ j=4\end{cases}$.

Then the wave field in the half-space surrounding the tunnel is composed of down-going waves and up-going waves.

$$\boldsymbol{u}(\boldsymbol{x})=\int_{-\infty}^{\infty}\mathrm{d}q\int_0^{\infty}\mathrm{d}p\sum_{j=1}^{4}\left(A_j(q,p)\phi_j^{-}(q,p,\boldsymbol{x})+\frac{2}{\pi}\phi_j^{+}(q,p,\boldsymbol{x})\sum_{m=0}^{\infty}B_{jm}(q)M_{jm}(p)\frac{1}{h_j}\right) \tag{3.2.26}$$

$$\boldsymbol{w}(\boldsymbol{x})=\int_{-\infty}^{\infty}\mathrm{d}q\int_0^{\infty}\mathrm{d}p\sum_{j=1}^{4}\left(\xi_jA_j(q,p)\phi_j^{-}(q,p,\boldsymbol{x})+\frac{2}{\pi}\phi_j^{+}(q,p,\boldsymbol{x})\sum_{m=0}^{\infty}\xi_jB_{jm}(q)M_{jm}(p)\frac{1}{h_j}\right) \tag{3.2.27}$$

$$p_{\mathrm{f}}(\boldsymbol{x})=\int_{-\infty}^{\infty}\mathrm{d}q\int_0^{\infty}\mathrm{d}p\sum_{j=3}^{4}\left(A_j(q,p)p_{\mathrm{f}}(\phi_j^{-}(q,p,\boldsymbol{x}))+\frac{2}{\pi}p_{\mathrm{f}}(\phi_j^{+}(q,p,\boldsymbol{x}))\sum_{m=0}^{\infty}B_{jm}(q)M_{jm}(p)\frac{1}{h_j}\right) \tag{3.2.28}$$

The traction-free and fully previous boundary conditions at the ground surface are

$$\sum_{j=1}^{4}\left(A_j(q,p)\boldsymbol{t}^{(\boldsymbol{e}_x)}(\phi_j^{-}(q,p,x=d))+\frac{2}{\pi}\boldsymbol{t}^{(\boldsymbol{e}_x)}(\phi_j^{+}(q,p,x=d))\sum_{m=0}^{\infty}B_{jm}(q)M_{jm}(p)\frac{1}{h_j}\right)=0 \tag{3.2.29}$$

$$\sum_{j=3}^{4}\left(A_j(q,p)p_{\mathrm{f}}(\phi_j^{-}(q,p,x=d))+\frac{2}{\pi}p_{\mathrm{f}}(\phi_j^{+}(q,p,x=d))\sum_{m=0}^{\infty}B_{jm}(q)M_{jm}(p)\frac{1}{h_j}\right)=0 \tag{3.2.30}$$

From Eqs. (3.2.29)–(3.2.30), the reflection coefficients are

$$A_j(q,p)=\frac{2}{\pi}\sum_{j'=1}^{4}R_{jj'}(q,p)\sum_{m'=0}^{\infty}B_{j'm'}(q)M_{j'm'}(p)\frac{1}{h_{j'}} \tag{3.2.31}$$

where $R_{jj'} = -\frac{r_{jj'}}{\Delta_j}$, m'=0, 1, … and j'=1, 2, 3.

The exponential functions can be expressed as

$$e^{-ih_s x}\sin py = \sum_{m=0}^{\infty}\varepsilon_m i^{-m} J_m(g_s r)\sin m\alpha_s \sin m\varphi \tag{3.2.32}$$

$$e^{-ih_{s,p1,p2}x}\cos py = \sum_{m=0}^{\infty}\varepsilon_m i^{-m} J_m(g_{s,p1,p2} r)\cos m\alpha_{s,p1,p2} \cos m\varphi \tag{3.2.33}$$

where $\varepsilon_0 = 1$ and $\varepsilon_m = 2$ for $m \geq 1$.

From Eqs. (3.2.32)–(3.2.33), down-going waves become

$$\phi_j^-(q,p,\boldsymbol{x}) = \sum_{m=0}^{\infty}\varepsilon_m \chi_{jm}^0(q,\boldsymbol{r}) M_{jm}(p) \tag{3.2.34}$$

Then the wave field in the half-space surrounding the tunnel is composed of outgoing waves and regular waves.

$$\boldsymbol{u}(\boldsymbol{r}) = \int_{-\infty}^{\infty}\mathrm{d}q\int_0^{\infty}\mathrm{d}p\sum_{j=1}^{4}A_j(q,p)\sum_{m=0}^{\infty}\varepsilon_m \chi_{jm}^0(q,\boldsymbol{r})M_{jm}(p) + \int_{-\infty}^{\infty}\mathrm{d}q\sum_{j=1}^{4}\sum_{m=0}^{\infty}B_{jm}(q)\chi_{jm}^+(q,\boldsymbol{r}) \tag{3.2.35}$$

$$\boldsymbol{w}(\boldsymbol{r}) = \int_{-\infty}^{\infty}\mathrm{d}q\int_0^{\infty}\mathrm{d}p\sum_{j=1}^{4}\xi_j A_j(q,p)\sum_{m=0}^{\infty}\varepsilon_m \chi_{jm}^0(q,\boldsymbol{r})M_{jm}(p) + \int_{-\infty}^{\infty}\mathrm{d}q\sum_{j=1}^{4}\sum_{m=0}^{\infty}\xi_j B_{jm}(q)\chi_{jm}^+(q,\boldsymbol{r}) \tag{3.2.36}$$

$$p_f(\boldsymbol{r}) = \int_{-\infty}^{\infty}\mathrm{d}q\int_0^{\infty}\mathrm{d}p\sum_{j=3}^{4}A_j(q,p)\sum_{m=0}^{\infty}\varepsilon_m p_f(\chi_{jm}^0(q,\boldsymbol{r}))M_{jm}(p) + \int_{-\infty}^{\infty}\mathrm{d}q\sum_{j=3}^{4}\sum_{m=0}^{\infty}B_{jm}(q)p_f(\chi_{jm}^+(q,\boldsymbol{r})) \tag{3.2.37}$$

Substituting the reflection coefficients into Eqs. (3.2.35)–(3.2.37), the wave field is rewritten as

$$\boldsymbol{u}(\boldsymbol{r}) = \int_{-\infty}^{\infty}\mathrm{d}q\sum_{j=1}^{4}\sum_{m=0}^{\infty}\left(\sum_{j'=1}^{4}\sum_{m'=0}^{\infty}Q_{jmj'm'}(q)B_{j'm'}(q)\chi_{jm}^0(q,\boldsymbol{r}) + B_{jm}(q)\chi_{jm}^+(q,\boldsymbol{r})\right) \tag{3.2.38}$$

$$\boldsymbol{w}(\boldsymbol{r}) = \int_{-\infty}^{\infty}\mathrm{d}q\sum_{j=1}^{4}\sum_{m=0}^{\infty}\left(\sum_{j'=1}^{4}\sum_{m'=0}^{\infty}Q_{jmj'm'}(q)\xi_j B_{j'm'}(q)\chi_{jm}^0(q,\boldsymbol{r}) + \xi_j B_{jm}(q)\chi_{jm}^+(q,\boldsymbol{r})\right) \tag{3.2.39}$$

$$p_f(\boldsymbol{r}) = \int_{-\infty}^{\infty}\mathrm{d}q\sum_{j=3}^{4}\sum_{m=0}^{\infty}\left(\sum_{j'=1}^{4}\sum_{m'=0}^{\infty}Q_{jmj'm'}(q)B_{j'm'}(q)p_f(\chi_{jm}^0(q,\boldsymbol{r})) + B_{jm}(q)p_f(\chi_{jm}^+(q,\boldsymbol{r}))\right) \tag{3.2.40}$$

where $Q_{jmj'm'}(q) = \varepsilon_m \frac{2}{\pi}\int_0^{\infty}R_{jj'}(q,p)M_{j'm'}(p)\frac{1}{h_{j'}}M_{jm}(p)\mathrm{d}p$.

3.2.2 Solution for an Elastic Cylinder Coupled with the Poroelastic Half-Space

In Section 3.1, the tunnel is modelled as a thin shell so only the displacement and the external force on the central line of the shell is considered. But modelling the tunnel as an elastic cylinder makes it possible to calculate the vibrations within the tunnel wall more accurately. Thus in this part the tunnel is modelled as an elastic cylinder with an inner radius a, an outer radius b as shown in Fig. 3.2.1.

The wave field in the tunnel is

$$\boldsymbol{u}_t(\boldsymbol{r})=\int_{-\infty}^{\infty}\mathrm{d}q\sum_{j=1}^{3}\sum_{m=0}^{\infty}(C_{jm}(q)\boldsymbol{\chi}_{jm}^{t0}(q,\boldsymbol{r})+D_{jm}(q)\boldsymbol{\chi}_{jm}^{t+}(q,\boldsymbol{r})) \tag{3.2.41}$$

where the superscript t on the variables indicates that the material parameters for the tunnel are used; C_{jm} and D_{jm} are the unknowns.

If a unit harmonic load $\boldsymbol{F}=\delta(\varphi-\pi)\delta(z)\mathrm{e}^{\mathrm{i}2\pi f_0 t}\boldsymbol{e}_r$ is applied at the tunnel invert, it can be expanded into Fourier series in the transformed domain as

$$\boldsymbol{t}^{e_r}(\boldsymbol{u}_t)\big|_{r=a}=\boldsymbol{e}_r\frac{1}{2\pi}\int_{-\infty}^{\infty}dq\mathrm{e}^{\mathrm{i}qz}\delta(\omega+2\pi f_0)\sum_{m=0}^{\infty}\frac{\varepsilon_m}{2\pi a}(-1)^m\cos m\varphi \tag{3.2.42}$$

The stresses at the inner surface of the tunnel are

$$\sum_{j=1}^{3}(C_{jm}(q)\boldsymbol{t}^{e_r}(\boldsymbol{\chi}_{jm}^{t0}(q,r=a)+D_{jm}(q)\boldsymbol{t}^{e_r}(\boldsymbol{\chi}_{jm}^{t+}(q,r=a))=\boldsymbol{e}_r\frac{1}{4\pi^2 a}\varepsilon_m(-1)^m\delta(\omega+2\pi f_0) \tag{3.2.43}$$

The stresses at the outer surface of the tunnel are

$$\begin{aligned}&\sum_{j=1}^{4}(\sum_{j'=1}^{4}\sum_{m'=0}^{\infty}Q_{jmj'm'}(q)B_{j'm'}(q)\boldsymbol{t}^{e_r}(\boldsymbol{\chi}_{jm}^{0}(q,r=b))+B_{jm}(q)\boldsymbol{t}^{e_r}(\boldsymbol{\chi}_{jm}^{+}(q,r=b)))=\\&\sum_{j=1}^{3}(C_{jm}(q)\boldsymbol{t}_1^{e_r}(\boldsymbol{\chi}_{jm}^{10}(q,r=b))+D_{jm}(q)\boldsymbol{t}_1^{e_r}(\boldsymbol{\chi}_{jm}^{1+}(q,r=b)))\end{aligned} \tag{3.2.44}$$

The displacements at the outer surface of the tunnel are

$$\begin{aligned}&\sum_{j=1}^{4}(\sum_{j'=1}^{4}\sum_{m'=0}^{\infty}Q_{jmj'm'}(q)B_{j'm'}(q)\boldsymbol{\chi}_{jm}^{0}(q,r=b)+B_{jm}(q)\boldsymbol{\chi}_{jm}^{+}(q,r=b))=\\&\sum_{j=1}^{3}(C_{jm}(q)\boldsymbol{\chi}_{jm}^{10}(q,r=b)+D_{jm}(q)\boldsymbol{\chi}_{jm}^{1+}(q,r=b))\end{aligned} \tag{3.2.45}$$

The concrete tunnel creates a fully impervious boundary at the tunnel-soil interface

$$\sum_{j=1}^{4}\sum_{m=0}^{\infty}(\sum_{j'=1}^{4}\sum_{m'=0}^{\infty}Q_{jmj'm'}(q)\xi_j B_{j'm'}(q)\chi_{jmr}^{0}(q,\boldsymbol{r})+\xi_j B_{jm}(q)\chi_{jmr}^{+}(q,\boldsymbol{r}))=0 \tag{3.2.46}$$

where $\chi_{jmr}^{0(+)}$ is the radial component of $\boldsymbol{\chi}_{jm}^{0(+)}$.

By solving Eqs. (3.2.44)–(3.2.46) we obtain the unknowns B_{jm} and D_{jm}, then they will be substituted into Eqs. (3.2.31) and (3.2.43) to get A_j and C_{jm}.

Numerical Results

The material parameters for the saturated soil and the tunnel are presented in Table 3.2.1, which are the same as those used in the reference[20]. The tunnel has an inner radius a=2.75 m and an outer radius b=3 m with a depth of d=10 m.

Table 3.2.1 Parameters for the poroelasic soil and the tunnel

Parameter	Value
Lamé constant for the solid skeleton, μ	1×10^8 N/m^2
Lamé constant for the solid skeleton, λ	2.333×10^8 N/m^2
Solid density, ρ_s	1816 kg/m^3
Water density, ρ_f	1000 kg/m^3
Porosity, n	0.4
Permeability, b	1×10^9 kg/(m^3·s)
Hysteretic material damping in the soil	0.1
Parameter for the compressibility of the soil particle, α	1.0
Parameter for the compressibility of the fluid, M	6.125×10^9 N/m^2
Lamé constant for the tunnel, μ_t	1.923×10^{10} N/m^2
Lamé constant for the tunnel, λ_t	2.885×10^{10} N/m^2
Density of the tunnel, ρ_t	2500 kg/m^3
Hysteretic material damping in the tunnel	0.03

Fig. 3.2.2 presents vertical displacement on the ground surface to a harmonic point load applied at the tunnel invert. Two observation points with coordinates x=10 m, y=0 m, z=0 m and x=10 m, y=10 m, z=0 m are chosen to show the variation of the transfer function with respect to the load frequency f_0. In Fig. 3.2.2, it is shown that the transfer function presents a behavior of oscillation with obvious peak and trough values. This is due to the fact that the vibrations at the ground surface are the superimposition of waves along different transmission paths, resulting in the wave interference phenomenon. In this case, constructive and destructive interferences lead to the peak and trough values in the displacement curves as shown in Fig. 3.2.2. Furthermore, it is found that at some frequencies, the displacements at the further observation point (x=10 m, y=10 m, z=0 m) are larger than those at the closer point (x=10 m, y=0 m, z=0 m) due to the spatial variation of the free-field responses.

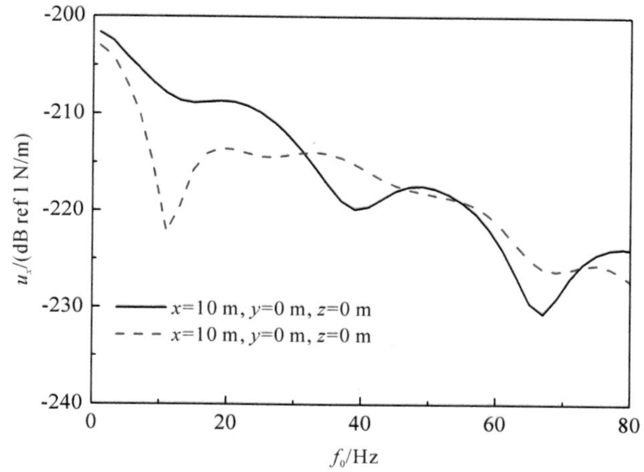

Fig. 3.2.2 Vertical displacement on the ground surface with coordinates x=10 m, y=0 m, z=0 m and x=10 m, y=10 m, z=0 m

The ground surface is fully pervious in the absence of structures, such as the embankment or the foundation, which means the pore pressure vanishes naturally. Thus Fig. 3.2.3 presents the pore pressure at the subsurface, which is 2 m below the ground surface. The coordinates of the two observation points are x=8 m, y=0 m, z=0 m and x=8 m, y=10 m, z=0 m. In Fig. 3.2.3, the pore pressure presents a behavior of oscillation like the displacement curves, which is attributed to the interference of longitudinal, transverse and Rayleigh waves. Similarly at some frequencies the pore pressure at the further point is larger than that at the closer point.

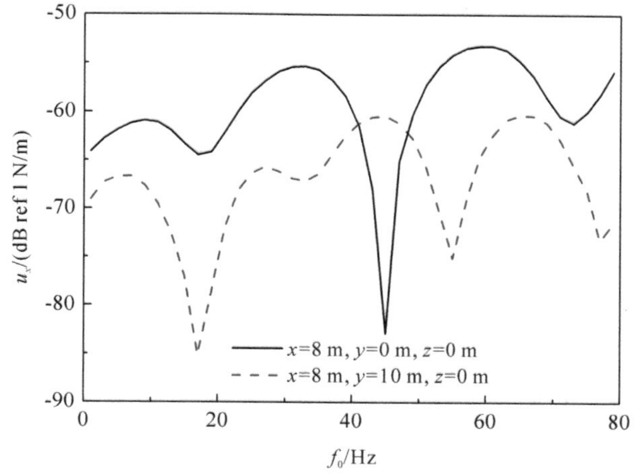

Fig. 3.2.3 Excess pore pressure at the subsurface (2m below the ground surface) with coordinates x=8 m, y=0 m, z=0 m and x=8 m, y=10 m, z=0 m

In this subsection, an analytical model for calculating vibrations from a tunnel embedded in a homogeneous half-space is proposed. Also, the present authors have improved it to account for the layered case. The procedure to couple the tunnel-soil model with the track structures is similar to that introduced in the previous Section 3.1. In the following context of this chapter, we will show some calculations for the project case of Xi'an metro lines.

3.2.3 Project case of Xi'an Metro Lines

The bell tower is a landmark of ancient buildings in Xi'an. It lies in the center of Xi'an city in China, as shown in Fig. 3.2.4. Xi'an metro line 2 passes through the bell tower by means of separate left and right tunnels. The central line of left or right tunnel is 17 m far from the terrace of the bell tower, and the tunnel vault depth is between 12–16 m. The central lines of two tunnels of line 6 are about 11–14 m far from the terrace of the bell tower and the tunnel vault depth is around 21 m. The relative positions between metro line 2, line 6 and the bell tower are shown in Fig. 3.2.5.

Fig. 3.2.4 The bell tower in Xi'an city

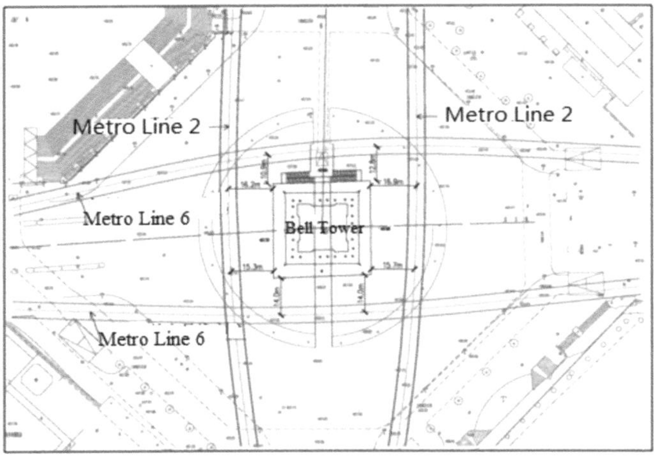

Fig. 3.2.5 Relative positions between line 2, line 6 and the bell tower in Xi'an city

To predict the ground vibrations generated by Xi'an metro lines, the analytical model developed previously will be used for modelling the moving trains, the track, the tunnel and the layered soil, aiming at providing some suggestions for mitigating the vibrations. Firstly we introduce the details of the designed train load, the railway track structure, the tunnel structure and the geological condition of the ground. Then the influence of the train velocity and the tunnel depth on surface vibrations will be investigated theoretically. Finally some measurements of the vibration at the bell tower pedestal done by Song et al.[21–22] will be used to validate the proposed analytical method. When modelling the vibrations from Xi'an metro lines by using the proposed analytical method, only line 2 is taken into account as the twin-tunnel case is not considered in the current method. The parameters of the designed train load, the track structure and the layered ground, which are listed in Tables 3.2.2–3.2.4, are chosen from the references [21–22]. Besides, the tunnel has Young's modulus 5×10^{10} N/m^2, Poisson's ratio 0.3 and a damping factor 0.05 with an outer diameter 6 m and a thickness 0.25 m.

Table 3.2.2 Parameters of the designed train load

Parameter	Value
Axle load, P_E	138.5 kN
Axle load, P_c	117.6 kN
Axle distance, w_a	2.2 m
Axle distance, w_b	12.6 m
Engine length, L_E	19.52 m
Carriage length, L_c	19.52 m
Total number of carriage, N_T	6

The designed train load consists of 3 engines (E) and 3 carriages (C) with a sequence of C-E-C-E-C-E. The axle load and the geometrical parameters for each carriage are given in Table. 3.2.2. The designed operating speed is 80 km/h. Here, the moving train loads are used, so the term on the right hand side of Eq. (3.1.28) should be replaced by

$$F=\sum_{n=1}^{N_\mathrm{T}}F_n(z-ct)\mathrm{e}^{\mathrm{i}2\pi f_0 t} \tag{3.2.47}$$

$$\begin{aligned}F_n(z-ct)=P_n[&\delta(z-ct+\sum_{s=0}^{n-1}L_\mathrm{s}+L_\mathrm{D})+\delta(z-ct+w_\mathrm{a}+\sum_{s=0}^{n-1}L_\mathrm{s}+L_\mathrm{D})\\&+\delta(z-ct+w_\mathrm{a}+w_\mathrm{b}+\sum_{s=0}^{n-1}L_\mathrm{s}+L_\mathrm{D})+\delta(z-ct+2w_\mathrm{a}+w_\mathrm{b}+\sum_{s=0}^{n-1}L_\mathrm{s}+L_\mathrm{D})]\end{aligned} \tag{3.2.48}$$

where P_n is the axle load of the engine or the carriage; L_s is the length of the engine or the carriage; L_D is the distance between the observation point and the first axle load of the moving trains in the z direction.

Table 3.2.3 Parameters of the floating slab tracks

Parameter	Value
Bending stiffness of the two rails, E_rI_r	6.72×10^6 Pa·m^4
Bending stiffness of the slab, E_sI_s	2.39×10^8 Pa·m^4
Mass per unit length of the two rails, m_r	121.8 kg/m
Mass per unit length of the slab, m_s	2571 kg/m
Stiffness of the rail pads, k_r	8.7×10^7 N/m^2
Stiffness of the slab bearing, k_s	1.14×10^7 N/m^2
Damping factors of the rail pads, c_r	1.6×10^5 N·s/m^2
Damping factors of the slab bearing, c_s	2.19×10^4 N·s/m^2

Table 3.2.4 Parameters of the layered saturated ground

Stratum No.	Ground layer	Thickness/m	Mass density/ (kg/m^3)	Poisson's ratio	Young's modulus/ MPa
1	Surface brick	0.4	2200	0.19	2079
2	Filling	8.2	1930	0.22	469
3	Clay	7.93	1820	0.23	224
4	Clay	4.03	1900	0.26	302
5	Clay	1.88	1980	0.25	353
6	Clay	4.95	2000	0.24	307

For the continuous floating slab track, two resonant frequencies exist, namely natural frequencies of the slab and the rail, which can be approximately calculated using the single degree system, i.e., $f_1=1/(2\pi)\sqrt{k_s/m_s}$ and $f_2=1/(2\pi)\sqrt{k_r/m_r}$. The substitution of the parameters in Table 3.2.3 into above two equations yields f_1=10.6 Hz and f_2=134.6 Hz. Resonance takes place when the load frequency reaches the natural frequencies of the track. So if we use the moving train loads with a fixed self-frequency f_0=10.6 Hz or 134.6 Hz, the maximum vibrations are obtained, which can be regarded as the upper bound of the vibrations. As f_0=134.6 Hz is beyond the frequency range we usually concern, f_0=10.6 Hz will be chosen as the load frequency in the following numerical results.

In Fig. 3.2.6, three tunnel depths d=14 m, 16 m and 18 m are selected to investigate the influence of the tunnel depth on surface vibrations for c=40 km/h. The vertical velocity at the ground surface directly above the tunnel invert shows a simple harmonic oscillation due to the fixed train load frequency, which gives the maximum vibrations. Because of the steady-state solution for the present case, the vertical velocity curves are plotted against the moving coordinate $z_0=z-ct$ that is usually used for the vibrations excitated by moving loads. It is observed that in Fig. 3.2.6, the vertical velocity is dramatically reduced by the increase of the tunnel depth. The decrease rate of the maximum vertical velocity is over 50% when increasing the tunnel depth from d=16 m to d=18 m. This is easy to be understood, because the propagating waves undergo more vibration circles from the vibration source to the observation point for a deeper tunnel, which means undergoing more damping effects. Therefore, increasing the tunnel depth is an effective way to reduce the vibrations caused by underground moving trains.

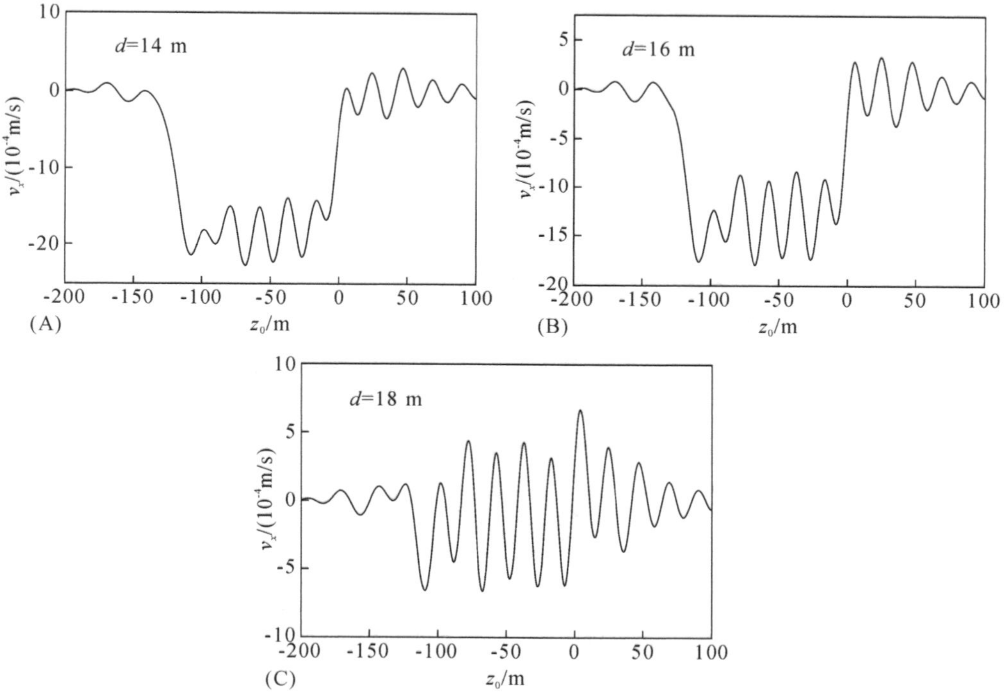

Fig. 3.2.6 Influence of the tunnel depth on the vertical velocity at the ground surface (c=40 km/h)

In Fig. 3.2.7, three train load velocities c=40 km/h, 80 km/h and 100 km/h are selected to investigate the influence of the load velocity on surface vibrations (d=18 m). As expected, the maximum vertical velocity decreases with the decreasing train load velocity, but the train load velocity has a smaller influence on surface vibrations than the tunnel depth. Thus increasing the tunnel depth is a better measure for controlling the ground-borne vibrations from underground railways. Besides increasing the tunnel depth and controlling the load velocity, comprehensive protection scheme is taken in the preliminary design to protect the bell tower from the negative impacts of the subway construction, such as using bored piles and rotary jet grouting piles. But it is obvious that the proposed analytical method is not capable of taking these factors into account due to the complex geometries.

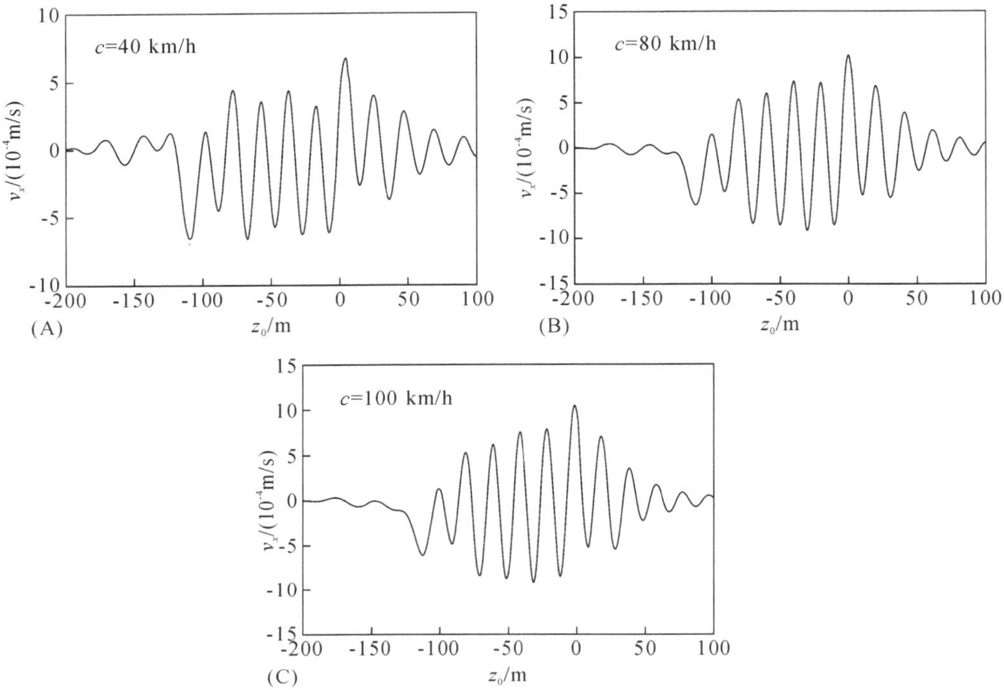

Fig. 3.2.7 Influence of the train speed on the vertical velocity at the ground surface (d=18 m)

In the end of this chapter we compare the calculated vibrations with the available measurement results done by Song et al.[21–22] to validate the proposed analytical model. 16 measurement points, whose locations are shown in Fig. 3.2.8 roughly, are selected on the bell tower pedestal. The time histories of the velocity responses are recorded for each test point in longitudinal, horizontal and vertical directions. Field test scenario can be seen in Fig. 3.2.9.

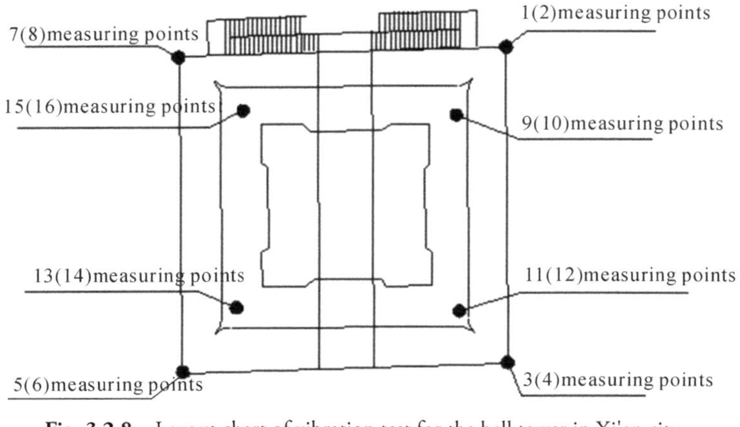

Fig. 3.2.8 Layout chart of vibration test for the bell tower in Xi'an city

Fig. 3.2.9 Vibration test site for the bell tower in Xi'an city

The maximum measured vibration velocities at points 1 and 3 on the bell tower pedestal, as shown in Fig. 3.2.8 are listed in Table 3.2.5. It can be seen that the vertical velocity of the bell tower pedestal has a maximum value compared to other two components. Since the calculations give the upper bound of the ground vibrations, the calculated values are larger than the measurement results, but of the same order, which proves the accuracy and the capability of current method to predict the vibrations from underground railways.

Table 3.2.5 Measured and calculated vibration velocities of the bell tower pedestal

Point No.	Working condition	Measurements Maximum velocity/ (10^{-4}m/s)			Calculations Maximum velocity/ (10^{-4}m/s)		
		L	H	V	L	H	V
1	Bidirectional operation of Line 2 at speed 70km/h	0.169	0.727	2.981	0.189	0.829	3.211
3	Bidirectional operation of Line 2 at speed 40km/h	0.075	0.663	2.380	0.089	0.716	2.717

[a]L, H and V denote the longitudinal, horizontal and vertical directions.

3.3 Conclusions

In this chapter, two analytical models for calculating vibrations from a tunnel embedded in the saturated poroelastic soil, i.e., a tunnel surrounded by a poroelastic full-space or half-space with a cylindrical cavity are introduced. These two models have the advantage of high computational efficiency and can be used to predict vibrations from underground railways by coupling them with the moving train and track elements.

The model of a tunnel embedded in a poroelatic full-space is suitable for the qualitative analysis and the evaluation of mitigation measures. From the numerical results in Section 3.1, it is found that the free-field responses generated by a constant moving load present resonance characteristics and the installation of floating slab tracks does not alter the critical velocity of the system; the natural frequency of the floating slab has significant influence on the soil displacement and if the load frequency is higher than the natural frequency of the slab, obvious isolation effect is achieved.

The model of a tunnel embedded in a poroelatic half-space is suitable for predicting the vibrations in a quantitative way. In Section 3.2, some parametric analysis about the train load velocity and the tunnel depth are performed for the ground vibrations from Xi'an metro lines by using the proposed model. The numerical results show that decreasing the train load velocity and increasing the tunnel depth are effective measures to reduce the surface vibrations. Furthermore, the comparison between the calculations and the measurements shows the accuracy and the capability of the current method to predict ground-borne vibrations.

Chapter 4
Problems for Vibrations of Foundations

This chapter introduces the procedures for solving the problem of a vibrating foundation in saturated ground using Biot's theory.

4.1 Vertical Vibrations of a Rigid Foundation Embedded in Saturated Soil

A simplified analytical method was used to investigate the vertical dynamic response of a rigid, massive, cylindrical foundation embedded in a poroelastic soil layer. The foundation was subjected to time-harmonic vertical loading and was perfectly bonded to the surrounding soil in the vertical direction. The behavior of the poroelastic soil is governed by Biot's poroelastodynamic theory and its governing equations are solved by means of the Hankel transform. The soil at the side of the foundation was simplified to be composed of a series of infinitesimally thin independent poroelastic layers, while the soil underlying the foundation base was modeled as a single-layered type of poroelastic soil based on rigid bedrock. The contact surface between the foundation base and the poroelastic soil was smooth and fully permeable. A computer code was developed to solve this interaction problem. Selected numerical results are presented to show the influence of nondimensional frequency of excitation, soil layer thickness, poroelastic material parameters, depth ratio and mass ratio of the foundation on the dynamic response of a rigid, massive, cylindrical foundation embedded in a poroelastic soil layer.

4.1.1 Governing Equations of Saturated Soil for Vertical Vibrations

The problem considered in this study is shown in Fig. 4.1.1, where a rigid cylindrical foundation with a radius r_0 and a mass m was embedded to a depth h in a poroelastic soil layer of thickness H. The foundation was subjected to a time-harmonic vertical loading $P(t) = P\mathrm{e}^{\mathrm{i}\omega t}$ acting along its axis of symmetry, where ω is the frequency of motion and $\sqrt{-1}$. A cylindrical coordinate system with its origin $o(r, \theta, z)$ located at the center of the foundation was defined such that the z-axis was perpendicular to the free surface of the soil layer, and the motion was axisymmetric with respect to the z-axis. The governing equations of a poroelastic medium for axisymmetric motion, in the absence of compression of the solid matrix and pore fluid, are given according to Eqs. (1.1.10), (1.1.20) as

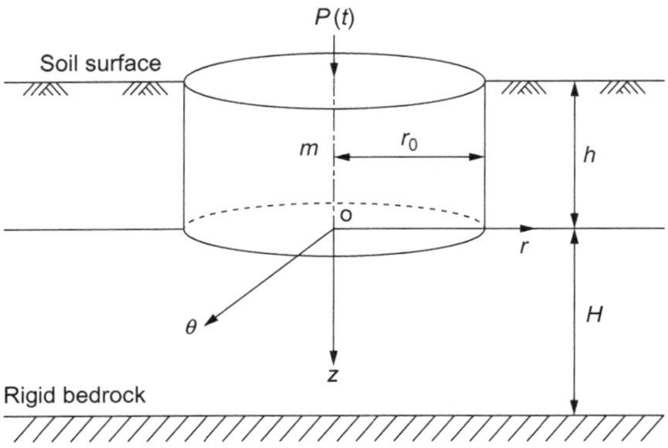

Fig. 4.1.1 Rigid cylindrical foundation embedded in a poroelastic soil layer

$$\frac{\partial \sigma_{rr}}{\partial r}+\frac{\sigma_{rr}-\sigma_{\theta\theta}}{r}+\frac{\partial \sigma_{rz}}{\partial z}-\frac{\partial p_{\mathrm{f}}}{\partial r}=\rho \ddot{u}_r+\rho_{\mathrm{f}}\ddot{w}_r \tag{4.1.1}$$

$$\frac{\partial \sigma_{rz}}{\partial r}+\frac{\sigma_{rz}}{r}+\frac{\partial \sigma_{zz}}{\partial z}-\frac{\partial p_{\mathrm{f}}}{\partial z}=\rho \ddot{u}_z+\rho_{\mathrm{f}}\ddot{w}_z \tag{4.1.2}$$

$$-\frac{\partial p_{\mathrm{f}}}{\partial r}=\rho_{\mathrm{f}}\ddot{u}_r+\frac{\rho_{\mathrm{f}}}{n}\ddot{w}_r+b\dot{w}_r \tag{4.1.3}$$

$$-\frac{\partial p_{\mathrm{f}}}{\partial z}=\rho_{\mathrm{f}}\ddot{u}_z+\frac{\rho_{\mathrm{f}}}{n}\ddot{w}_z+b\dot{w}_z \tag{4.1.4}$$

The constitutive relations for a homogeneous poroelastic material can be expressed according to Eq. (1.1.21) as

$$\sigma_{rr}=\lambda e+2\mu\frac{\partial u_r}{\partial r} \tag{4.1.5}$$

$$\sigma_{zz}=\lambda e+2\mu\frac{\partial u_z}{\partial z} \tag{4.1.6}$$

$$\sigma_{\theta\theta}=\lambda e+2\mu\frac{u_r}{r} \tag{4.1.7}$$

$$\sigma_{rz}=\mu\left(\frac{\partial u_r}{\partial z}+\frac{\partial u_z}{\partial r}\right) \tag{4.1.8}$$

$$\frac{\partial \dot{u}_r}{\partial r}+\frac{\dot{u}_r}{r}+\frac{\partial \dot{u}_z}{\partial z}+\frac{\partial \dot{w}_r}{\partial r}+\frac{\dot{w}_r}{r}+\frac{\partial \dot{w}_z}{\partial z}=0 \tag{4.1.9}$$

The motion under consideration was assumed to be time-harmonic with the factor $e^{i\omega t}$, which was henceforth suppressed from all expressions for brevity. It was convenient to use nondimensional forms of length parameters including the coordinates by selecting the radius of the foundation as a unit of length. Stresses and pore pressure can also be normalized into the nondimensional forms, with respect to the shear modulus μ of the bulk material. All variables were replaced by the nondimensional quantities, but the previous notations were used for convenience. In addition, the following nondimensional frequency and material properties were introduced:

$$a_0 = \sqrt{\frac{\rho}{\mu}} r_0 \omega, \quad \lambda^* = \frac{\lambda}{\mu}, \quad \rho^* = \frac{\rho_f}{\rho}, \quad b^* = \frac{b r_0}{\sqrt{\rho \mu}}, \quad M^* = M/\mu \tag{4.1.10}$$

4.1.2 Boundary Conditions for Vertical Vibrations

For the vibration problem of a rigid foundation perfectly bonded to the surrounding soil in the vertical direction, as shown in Fig. 4.1.1, it was assumed that the contact surface between the foundation base and the poroelastic soil was smooth and fully permeable.

The boundary conditions on $z = 0$ and $z = H/r_0$ can then be written in nondimensional quantities as follows:

$$u_z(r, 0) = U_0, \ 0 \le r \le 1 \tag{4.1.11}$$

$$\sigma_{zz}(r, 0) = 0, \ 1 \le r \le \infty \tag{4.1.12}$$

$$\sigma_{rz}(r, 0) = 0, \ 0 \le r \le \infty \tag{4.1.13}$$

$$p_f(r, 0) = 0, \ 0 \le r \le \infty \tag{4.1.14}$$

$$u_z(r, H/r_0) = 0, \ 0 \le r \le 1 \tag{4.1.15}$$

$$u_r(r, H/r_0) = 0, \ 0 \le r \le 1 \tag{4.1.16}$$

$$w_z(r, H/r_0) = 0, \ 0 \le r \le 1 \tag{4.1.17}$$

where $U_0 = U/r_0$ with U denoting the vertical displacement of the rigid foundation.

4.1.3 A Simplified Analytical Method for Vibrations of Foundations

Equation of Dynamic Equilibrium for the Rigid Foundation

In addition to those previously defined, the following notations were also introduced in this study: $R_b(t) = R_b e^{i\omega t}$ is the vertical dynamic reaction at the foundation base; $R_s(t) = R_s e^{i\omega t}$ is the vertical dynamic reaction along the side of the foundation. The equation of the dynamic equilibrium for the foundation is

$$m\ddot{U}(t) = P(t) - R_b(t) - R_s(t) \tag{4.1.18}$$

where $U(t) = U\mathrm{e}^{\mathrm{i}\omega t}$.

Because of the difficulty in obtaining an exact analytical solution for a dynamic response of embedded foundations, a simplified analytical method proposed by Baranov[23] was extended and used by Novak et al.[24–26] to study the vertical, coupled horizontal and rocking, torsional and coupled vibrations of a rigid foundation embedded in an ideal elastic soil. In the Baranov-Novak method, the soil was divided into two independent parts and it was assumed that the soil reaction at the base of the foundation was equal to that of a foundation resting on the surface of the soil, while the lateral soil reaction was evaluated independently. In this paper, several similar assumptions were introduced: (I) the soil at the side of the foundation is composed of a series of infinitesimally thin independent poroelastic layers. In each infinitesimally thin poroelastic layer, the influence of the gradient of σ_{zz} and p_{f} in the vertical direction and the influence of radial displacement on $R_{\mathrm{s}}(t)$ are neglected. (II) The dynamic reaction at the base of the foundation is independent of depth of embedment.

Dynamic Reaction at the Base of the Foundation

Under assumption (II), the dynamic reaction at the base of the foundation can be derived from a dynamic interaction between a surface foundation and a single-layered poroelastic soil. The governing partial differential equations of poroelastic soil for axisymmetric motion can be solved by applying a Hankel integral transform with respect to the radial coordinate r. The μth-order Hankel integral transform of a function $f(r, z)$ with respect to r and its inverse transform are defined by Sneddon[27] as

$$\tilde{f}^{\mu}(\xi, z) = \int_0^{\infty} rf(r, z)J_{\mu}(\xi r)\mathrm{d}r \tag{4.1.19}$$

$$f^{\mu}(r, z) = \int_0^{\infty} \xi\tilde{f}(\xi, z)J_{\mu}(\xi r)\mathrm{d}\xi \tag{4.1.20}$$

where J_{μ} is the Bessel function of the first kind of the μth order, and ξ is the Hankel transform parameter.

The following general solutions can be obtained for a zero-order Hankel transform of u_z, w_z, σ_{zz}, and p_{f}, and the first-order Hankel transform of σ_{rz} and u_r, respectively:

$$\tilde{p}_{\mathrm{f}}^{0} = \chi_1 A_1 \mathrm{e}^{-qz} + \chi_1 B_1 \mathrm{e}^{qz} + A_2 \mathrm{e}^{-\xi z} + B_2 \mathrm{e}^{\xi z} \tag{4.1.21}$$

$$\tilde{u}_r^{1} = \chi_2 A_1 \mathrm{e}^{-qz} + \chi_2 B_1 \mathrm{e}^{qz} + \chi_3 A_2 \mathrm{e}^{-\xi z} + \chi_3 B_2 \mathrm{e}^{-\xi z} + A_3 \mathrm{e}^{-sz} + B_3 \mathrm{e}^{sz} \tag{4.1.22}$$

$$\tilde{u}_z^{0} = \chi_4 A_1 \mathrm{e}^{-qz} - \chi_4 B_1 \mathrm{e}^{qz} + \chi_3 A_2 \mathrm{e}^{-\xi z} - \chi_3 B_2 \mathrm{e}^{-\xi z} + \chi_5 A_3 \mathrm{e}^{-sz} - \chi_5 B_3 \mathrm{e}^{sz} \tag{4.1.23}$$

$$\tilde{w}_z^{0} = \kappa_1 \left[\tilde{u}_z^{0} - \frac{1}{\rho^* {a_0}^2}\left(-q\chi_1 A_1 \mathrm{e}^{-qz} + q\chi_1 B_1 \mathrm{e}^{qz} - \xi A_2 \mathrm{e}^{-\xi z} + \xi B_2 \mathrm{e}^{\xi z}\right)\right] \tag{4.1.24}$$

$$\tilde{\sigma}_{zz}^{0} = \gamma_1 A_1 \mathrm{e}^{-qz} + \gamma_1 B_1 \mathrm{e}^{qz} - 2\xi\chi_3 A_2 \mathrm{e}^{-\xi z} - 2\xi\chi_3 B_2 \mathrm{e}^{-\xi z} - 2\xi A_3 \mathrm{e}^{-sz} - 2\xi B_3 \mathrm{e}^{sz} \tag{4.1.25}$$

$$\tilde{\sigma}_{rz}^{1} = \gamma_2 A_1 \mathrm{e}^{-qz} - \gamma_2 B_1 \mathrm{e}^{qz} - 2\xi\chi_3 A_2 \mathrm{e}^{-\xi z} + 2\xi\chi_3 B_2 \mathrm{e}^{\xi z} + \gamma_3 A_3 \mathrm{e}^{-sz} - \gamma_3 B_3 \mathrm{e}^{sz} \tag{4.1.26}$$

where A_1, A_2, A_3, B_1, B_2, and B_3 are arbitrary functions to be determined from the boundary and continuity conditions of the given problem. The variables κ_1, q, s, χ_i (i = 1, 2, 3, 4, 5), and γ_j (j = 1, 2, 3) appearing in the preceding equations are known functions of the material properties and the frequency of excitation, and are given in Appendix C.

By applying the Hankel transform to Eqs. (4.1.12)–(4.1.17), together with Eqs. (4.1.21)–(4.1.26), the boundary conditions can be rewritten in a matrix form:

$$\boldsymbol{Q}\cdot\boldsymbol{a}=\boldsymbol{f} \tag{4.1.27}$$

where $\boldsymbol{a}^{\mathrm{T}}=\{A_1, A_2, A_3, B_1, B_2, B_3\}$; $\boldsymbol{f}^{\mathrm{T}}=\{\tilde{\sigma}_{zz}^0(\xi, 0), 0, 0, 0, 0, 0\}$; and $\boldsymbol{Q}=\begin{bmatrix} \boldsymbol{Q}_{11} & \boldsymbol{Q}_{21} \\ \boldsymbol{Q}_{12} & \boldsymbol{Q}_{22} \end{bmatrix}$ denotes a 6 × 6 matrix whose elements $\boldsymbol{Q}_{11}$, $\boldsymbol{Q}_{12}$, $\boldsymbol{Q}_{21}$, and $\boldsymbol{Q}_{22}$ are functions of ξ and are given in Appendix D.

The solution of Eq. (4.1.27) is

$$A_1=\frac{\varphi_{11}^*}{|\boldsymbol{\varphi}|}\tilde{\sigma}_{zz}^0(\xi,0),\quad A_2=\frac{\varphi_{12}^*}{|\boldsymbol{\varphi}|}\tilde{\sigma}_{zz}^0(\xi,0),\quad A_3=\frac{\varphi_{13}^*}{|\boldsymbol{\varphi}|}\tilde{\sigma}_{zz}^0(\xi,0) \tag{4.1.28}$$

$$B_1=\frac{\psi_{11}^*}{|\boldsymbol{\psi}|}\tilde{\sigma}_{zz}^0(\xi,0),\quad B_2=\frac{\psi_{12}^*}{|\boldsymbol{\psi}|}\tilde{\sigma}_{zz}^0(\xi,0),\quad B_3=\frac{\psi_{13}^*}{|\boldsymbol{\psi}|}\tilde{\sigma}_{zz}^0(\xi,0) \tag{4.1.29}$$

where $\boldsymbol{\varphi}=\boldsymbol{Q}_{11}-\boldsymbol{Q}_{12}\boldsymbol{Q}_{22}^{-1}\boldsymbol{Q}_{21}$; $\boldsymbol{\psi}=\boldsymbol{Q}_{12}-\boldsymbol{Q}_{11}\boldsymbol{Q}_{21}^{-1}\boldsymbol{Q}_{22}$; $|\boldsymbol{\varphi}|$ and $|\boldsymbol{\psi}|$ denote the determinants of $\boldsymbol{\varphi}$ and $\boldsymbol{\psi}$, respectively; φ_{11}^*, φ_{12}^*, φ_{13}^* and ψ_{11}^*, ψ_{12}^*, ψ_{13}^* denote the algebraic complements of φ_{11}, φ_{12}, φ_{13} and ψ_{11}, ψ_{12}, ψ_{13}, corresponding to φ and ψ, respectively.

Substitution of Eqs. (4.1.28), (4.1.29) into Eq. (4.1.23) results in the following relationship:

$$\tilde{u}_z^0(\xi,0)=p(\xi)\tilde{\sigma}_{zz}^0(\xi,0) \tag{4.1.30}$$

where $p(\xi)=\dfrac{\chi_4\varphi_{11}^*+\chi_3\varphi_{12}^*+\chi_5\varphi_{13}^*}{|\boldsymbol{\varphi}|}-\dfrac{\chi_4\psi_{11}^*+\chi_3\psi_{12}^*+\chi_5\psi_{13}^*}{|\boldsymbol{\psi}|}$.

Then substituting Eq. (4.1.30) into Eq. (4.1.11) and expressing $\tilde{\sigma}_{zz}^0(r, 0)$ in terms of its Hankel transform, a set of dual integral equations describing the vertical vibration of a rigid foundation embedded in a poroelastic soil layer are finally derived as follows:

$$\int_0^{\infty}\xi^{-1}[1+H(\xi)]B(\xi)J_0(\xi r)\mathrm{d}\xi=-\frac{U_0}{1-\nu},\quad 0\le r\le 1 \tag{4.1.31}$$

$$\int_0^{\infty}B(\xi)J_0(\xi r)\mathrm{d}\xi=0,\quad r>1 \tag{4.1.32}$$

where $B(\xi)=\xi\,\tilde{\sigma}_{zz}^0(\xi, 0)$, $H(\xi)=-\dfrac{\xi p(\xi)}{1-\nu}-1$ and ν is Poisson's ratio of the solid matrix. It can be proved that $\lim\limits_{\xi\to\infty}\xi p(\xi)=-(1-\nu)$, i.e., $\lim\limits_{\xi\to\infty}H(\xi)=0$.

The dual integral equations can be reduced to a Fredholm integral equation of the second kind by employing the following integral representation:

$$B(\xi) = -\frac{2\xi U_0}{\pi(1-\nu)}\int_0^1 \theta(x)\cos(\xi x)\mathrm{d}x \tag{4.1.33}$$

Substituting Eq. (4.1.33) into Eqs. (4.1.31), (4.1.32), it can be seen that Eq. (4.1.32) is automatically satisfied, while Eq. (4.1.31) is equivalent to the following integral equation of the Fredholm type:

$$\theta(x) + \frac{1}{\pi}\int_0^1 F(x,\tau)\theta(\tau)\mathrm{d}\tau = 1 \tag{4.1.34}$$

where $F(x, \tau)$ is the kernel function, defined as

$$F(x,\tau) = 2\int_0^{\infty} H(\xi)\cos(\xi x)\cos(\xi\tau)\mathrm{d}\xi \tag{4.1.35}$$

By applying the trapezoidal rule to the finite integral appearing in Eq. (4.1.34), the integral equation derived previously can be discretized into a system of algebraic equations and subsequently solved by a computer program following this procedure. In deriving Eq. (4.1.34) the following relationship has been invoked:

$$\frac{R_b}{U} = \frac{4\mu r_0}{1-\nu}\int_0^1 \theta(x)\mathrm{d}x = \frac{4\mu r_0}{1-\nu}\frac{1}{f_1 + \mathrm{i}f_2} \tag{4.1.36}$$

where $f_1=\mathrm{Re}\left(\frac{1}{\int_0^1 \theta(x)\mathrm{d}x}\right)$ and $f_2 = \mathrm{Im}\left(\frac{1}{\int_0^1 \theta(x)\mathrm{d}x}\right)$ are the nondimensional dynamic compliance coefficients of a rigid foundation resting on a poroelastic soil surface and can be obtained by direct integration of the numerical solution of $\theta(x)$ from Eq. (4.1.34).

Dynamic Reaction Along the Side of the Foundation

Following assumption (I), Eqs. (4.1.2), (4.1.4), (4.1.8) can be rewritten in terms of nondimensional displacement components u_z and w_z for the time-harmonic dynamic case as

$$\frac{\partial^2 u_z}{\partial r^2} + \frac{1}{r}\frac{\partial u_z}{\partial r} = -{a_0}^2 u_z - \rho^* {a_0}^2 w_z \tag{4.1.37}$$

$$-n\rho^* a_0 u_z + (\mathrm{i}nb^* - \rho^* a_0)w_z = 0 \tag{4.1.38}$$

$$\sigma_{rz} = \frac{\partial u_z}{\partial r} \tag{4.1.39}$$

Combining Eqs. (4.1.37), (4.1.38) yields

$$r^2\frac{\partial^2 u_z}{\partial r^2}+\frac{1}{r}\frac{\partial u_z}{\partial r}+d^2r^2u_z=0 \tag{4.1.40}$$

The general solution of Eq. (4.1.40) is

$$u_z=C_1H_0^{(1)}(dr)+C_2H_0^{(2)}(dr) \tag{4.1.41}$$

where C_1 and C_2 are arbitrary functions; $H_0^{(1)}$ and $H_0^{(2)}$ denote modified Bessel functions of the first and second kind of the zero-order, respectively. Since the soil was assumed to be infinite in the radial direction, C_1 = 0 to guarantee boundary conditions as $r\to\infty$. The variable d is given in Appendix C. By virtue of Eqs. (4.1.41), (4.1.39), the nondimensional shear stress can be expressed as

$$\sigma_{rz}=-C_2\,dH_1^{(2)}(dr) \tag{4.1.42}$$

where $H_1^{(2)}$[28] is the modified Bessel functions of the second kind of the first order.

In view of Eqs. (4.1.11), (4.1.41), (4.1.42), the expression of σ_{rz} for r=1, denoted by σ_{1z}, is derived as

$$\sigma_{1z}=-\frac{dH_1^{(2)}(d)}{H_0^{(2)}(d)}U_0 \tag{4.1.43}$$

The dynamic reaction R_s can be expressed as

$$R_s=-Gr_0\int_0^l 2\pi r_0\sigma_{1z}\mathrm{d}z=Gr_0(S_1+\mathrm{i}S_2)U \tag{4.1.44}$$

where $S_1=\mathrm{Re}\left[2\pi l\frac{dH_1^{(2)}(d)}{H_0^{(2)}(d)}\right]$; $S_1=\mathrm{Im}\left[2\pi l\frac{dH_1^{(2)}(d)}{H_0^{(2)}(d)}\right]$ and $l=h/r_0$.

Dynamic Interaction Between the Embedded Foundation and the Poroelastic Soil Layer

Substitution of Eqs. (4.1.36), (4.1.44) into Eq. (4.1.18) yields

$$\frac{\rho r_0{}^2B_z}{G}\ddot{U}(t)+c\dot{U}(t)+kU(t)=\frac{1-\nu}{4Gr_0}P(t) \tag{4.1.45}$$

where B_z is the mass ratio of the foundation, and is defined as Lysmer and Richart[29]:

$$B_z=\frac{1-\nu}{4}\frac{m}{\rho r_0{}^3} \tag{4.1.46}$$

and

$$k=\frac{f_1}{f_1{}^2+f_2{}^2}+\frac{1-\nu}{4}S_1,\quad c=\left(\frac{-f_2}{f_1{}^2+f_2{}^2}+\frac{1-\nu}{4}S_2\right)/a_0 \tag{4.1.47}$$

The vertical displacement amplitude of the foundation U_{max} is given in Lysmer and Richert[29] as

$$U_{max} = \frac{P}{Gr_0}\frac{1-\nu}{4} \cdot M \tag{4.1.48}$$

where M is the dynamic response factor and is expressed as

$$M = \frac{1}{\sqrt{(k - B_z a_0{}^2)^2 + (ca_0)^2}} \tag{4.1.49}$$

The nondimensional vertical dynamic impedance, denoted by K_v, is given by

$$K_v = k + ica_0 \tag{4.1.50}$$

4.1.4 Numerical Results and Conclusions

A computer code based on the solution scheme described in the preceding sections has been developed to study the vertical dynamic response of a rigid cylindrical foundation embedded in a poroelastic soil layer. The response factor and dynamic impedance can be calculated by Eqs. (4.1.49), (4.1.50). To focus on studying the vertical response of embedded foundations in a poroelastic soil layer under dynamic excitation, the analysis presented in this paper was performed in the nondimensional frequency range of interest of $0.1 \le a_0 \le 5$.

Comparisons with Published Solutions

The accuracy of the present solution scheme is verified by comparing with well-known solutions obtained from other approaches, for both the elastodynamic interaction problem and the poroelastodynamic interaction problem. The numerical solutions corresponding to the elastic medium were based on the reduction of a poroelastic material to an ideal case by setting negligibly small values of the poroelastic parameters (ρ^* and b^* were set to 10^{-5}). In addition, solutions corresponding to the half-space were selected by $H/r_0 = 50$. (This depth was considered to be large enough to satisfy the half-space assumption.)

Fig. 4.1.2 shows a comparison of the dynamic response factor M of a rigid cylindrical foundation embedded in an elastic half-space. A mass ratio $B_z = 4$ and three values of depth ratio l ($l = 0, 0.5, 1.0$) were used in the comparison. The solutions obtained from the present scheme were then compared with the reported solution by Lysmer and Kuhlemeyer[30] using the same material parameters. They reached the solution by using a finite element method, and the response factor for surface foundations ($l = 0$) was calculated in a nondimensional frequency range of $0 < a_0 \le 2$, whereas for embedded foundations ($l = 0.5, 1.0$) the curves were only given in the range of $0.25 \le a_0 \le 1.5$. It can be seen that a very close agreement was obtained between the two solutions.

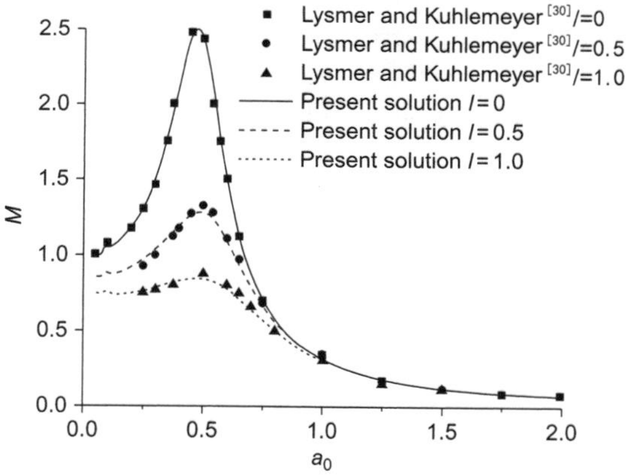

Fig. 4.1.2 Comparison of response factor for a rigid foundation embedded in an elastic half-space

Fig. 4.1.3 shows a comparison of nondimensional vertical dynamic impedance K_v for a rigid cylindrical foundation (l = 0.5, 1.0) embedded in an elastic half-space (ν = 0.25) obtained from the present scheme and the solutions given by Apsel and Luco[31]. The comparison was conducted in the range of $0.1 \le a_0 \le 5$. Apsel and Luco modeled the soil as a viscoelastic material in the half-space and the dynamic interaction problem was solved by a boundary integral equation approach. As shown in Fig. 4.1.3, the imaginary part of the dynamic impedance agrees well with the solutions of Apsel and Luco. For the real part, some differences occurred: in the lower frequency range, the present solution was less than the reported value, and the difference decreased as a_0 increased: about 8% difference when a_0 = 0.1, and 4% when a_0 = 0.25; at higher frequencies, the present solution was larger than the reported data, but the variation is quite small (maximum of 2%). Those differences are due to the fact that the present solutions correspond to an elastic soil with zero material damping while Apsel and Luco's solutions included a small amount of material damping.

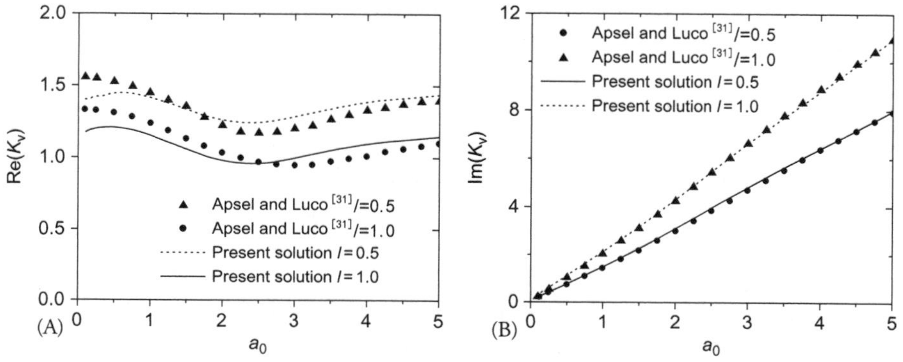

Fig. 4.1.3 Comparison of nondimensional vertical dynamic impedance for a rigid foundation embedded in an elastic half-space

Fig. 4.1.4 compares the nondimensional dynamic compliance coefficients f_1 and f_2 for a rigid foundation resting on the surface of a poroelastic half-space (l = 0) with those given by Zeng and Rajapakse[32]. The material properties of the poroelastic soil are identical to those of Zeng and Rajapakse. Again, excellent agreement was obtained between the two solutions, as presented in Fig. 4.1.4.

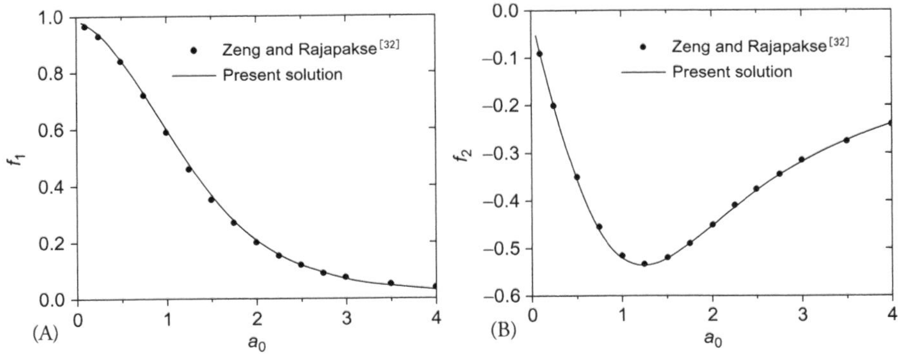

Fig. 4.1.4 Comparison of nondimensional vertical dynamic compliance coefficients for a rigid foundation resting on the surface of a poroelastic half-space

These independent comparisons therefore confirm the accuracy of the present solution scheme.

Numerical Results for Poroelastic Soil Layer

In the numerical study, some results for the vertical dynamic impedance and dynamic response factor of a rigid, massive, cylindrical foundation embedded in a poroelastic soil layer are presented to show the effect of nondimensional frequency of excitation, the thickness of the soil layer, poroelastic material parameters, depth ratio and mass ratio of the foundation. The following nondimensional material parameters were used for all numerical results presented in this section: λ^* = 1.0, ρ^* = 0.53, and n = 0.4. The parameter b^* was chosen to be 10 except in the studies considering the effect of b^*.

Effect of the Thickness of the Soil Layer

The effect of the thickness of the soil layer was investigated first. In this study, five values of the thickness of the soil layer were considered, H/r_0 = 2, 5, 10, 15, 20, and the results corresponding to a poroelastic half-space ($H/r_0 \to \infty$) are also shown for comparison.

Figs. 4.1.5 and 4.1.6 present the effect of the thickness of the soil layer on nondimensional dynamic impedance for a surface foundation (l = 0) and an embedded foundation with a depth ratio l = 0.5, respectively. It can be seen that both real and imaginary parts of the dynamic impedance show significant fluctuations with the increase of nondimensional frequency a_0 when the thickness of the soil layer was small. The variation of dynamic impedance with a_0 is quite similar for surface foundations and embedded foundations, although both real and imaginary parts of the dynamic impedance were smaller for surface foundations. Due to the role of soil-layer resonant phenomena, the real part of the dynamic

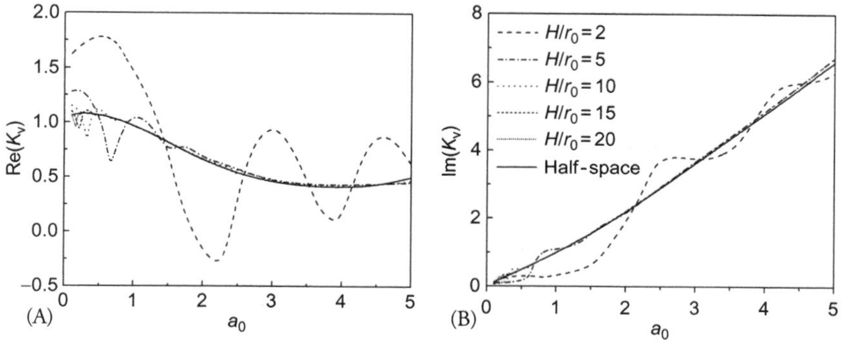

Fig. 4.1.5 Effect of the thickness of the soil layer on vertical dynamic impedance for a surface foundation

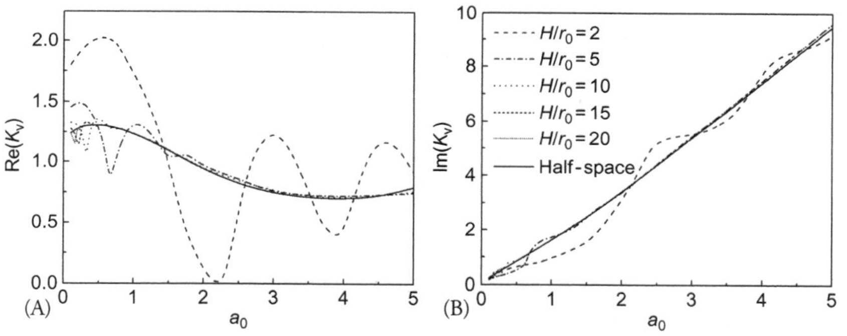

Fig. 4.1.6 Effect of the thickness of the soil layer on vertical dynamic impedance for an embedded foundation (l = 0.5)

impedance exhibited valleys at specific frequencies, and both the valley values and resonant frequencies decreased with the increase in H/r_0. On the other hand, the imaginary part of the dynamic impedance was not a linearly increasing function of a_0 as in the case of the half-space, but the curves displayed oscillatory variations with a_0. However, as the thickness of the soil layer increased, the fluctuations became less pronounced and the curves tended to be stable and gradually approached the results from the half-space. For the deepest soil layer considered, H/r_0 = 20, the effect of the thickness of the soil layer can be negligible and the fluctuation was only observed in the frequency range of $a_0 \leq 0.2$, which implies that for a very thick soil layer ($H/r_0 \geq 20$), the solution of the dynamic impedance of a half-space can be used as an approximation without causing much significant error.

Fig. 4.1.7 presents the effect of the thickness of the soil layer on the dynamic response factor for a massless foundation (B_z = 0) and for a foundation with B_z = 2. Only the result for the case of l = 0.5 is presented since similar trends were obtained for different values of l. As shown in Fig. 4.1.7A, unlike the half-space, it was found that the response factor for a massless foundation embedded in a soil layer was not a smooth decreasing function of nondimensional frequency, but exhibited peaks related to the natural frequencies of the soil layer. In the case of a massive foundation, as shown in

Fig. 4.1.7B, the peak values of the response factor for a soil layer were greater than those for the half-space case. It can be further seen that for a massive foundation, the thickness of the soil layer affected the response factor in a relatively narrow range of frequency, when compared to the massless case. The numerical results presented in Fig. 4.1.7 clearly indicate that the solutions of the half-space cannot be used to approximate the response of soil layer except for $H/r_0 \geq 20$ or at higher frequencies ($a_0 \geq 3$ for a massless foundation and $a_0 \geq 1.5$ for a massive foundation).

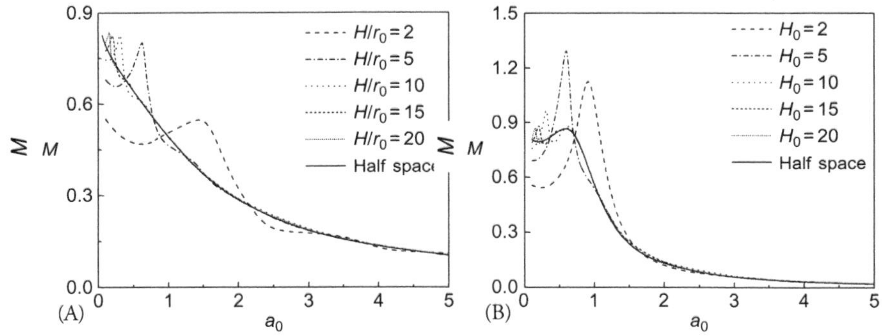

Fig. 4.1.7 Effect of the thickness of the soil layer on vertical response factor for an embedded foundation ($l = 0.5$). (A) $B_z = 0$ and (B) $B_z = 2$

Effect of Depth Ratio of the Foundation

Most real foundations are embedded in the soil to a certain extent. Embedment may result in an increase in the stiffness of the foundation-soil system due to the increase in the contact area between the foundation and the soil. The increase, however, depends on the type and the degree of contact with the surrounding soil.

In order to study the effect of embedment, three values of depth ratio were chosen with $l = 0$, 0.25, and 0.5 and the corresponding results are shown in Figs. 4.1.8–4.1.11, for two values of the thickness of the soil layer: $H/r_0 = 2$ and 20.

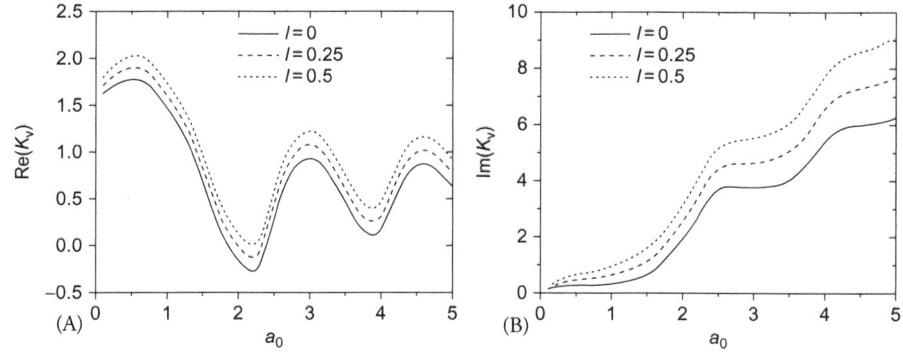

Fig. 4.1.8 Effect of depth ratio on dynamic impedance for a soil layer with $H/r_0 = 2$

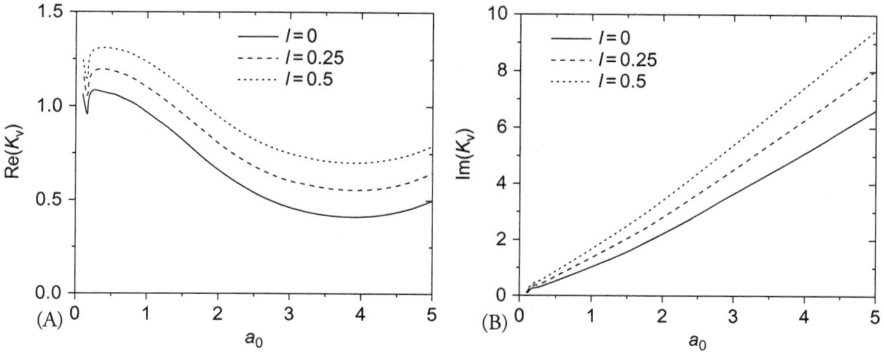

Fig. 4.1.9 Effect of depth ratio on dynamic impedance for a soil layer with H/r_0 = 20

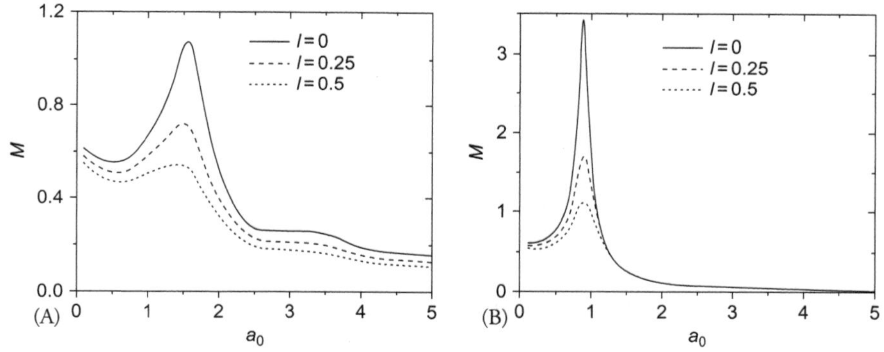

Fig. 4.1.10 Effect of depth ratio on the dynamic response factor for a soil layer with H/r_0 = 2. (A) B_z = 0 and (B) B_z = 2

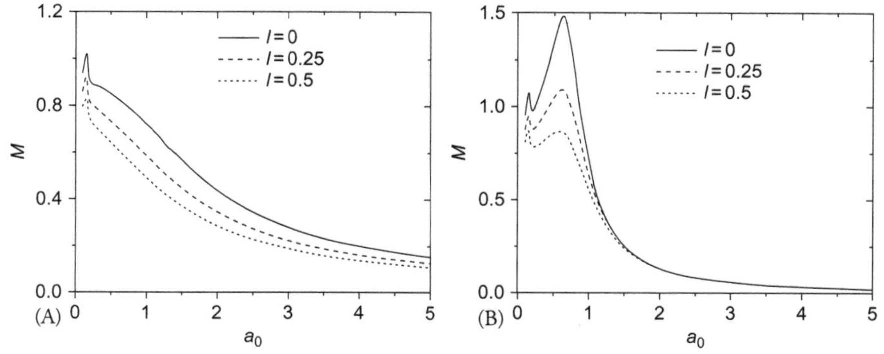

Fig. 4.1.11 Effect of depth ratio on the dynamic response factor for a soil layer with H/r_0 = 20. (A) B_z = 0 and (B) B_z = 2

The effect of depth ratio on vertical dynamic impedance is shown in Figs. 4.1.8 and 4.1.9. It is evident that the dynamic impedance depends significantly on depth ratio, for both real and imaginary parts of the dynamic impedance increasing with increasing depth ratio for both shallow and thick soil layers. However, the variations of the dynamic impedance with the frequency a_0 showed a similar trend for all the values of depth ratio considered. For the real part of the dynamic impedance, the curves yielded minimum values at some resonant frequencies. In the case of the imaginary parts, the curves of a shallow soil layer showed fluctuations with a_0, while in a thick soil layer with $H/r_0 = 20$, as shown in Fig. 4.1.9, the curves showed a nearly linear increase with a_0, and the larger the depth ratio, the larger the increase rate.

Figs. 4.1.10 and 4.1.11 show the effect of depth ratio on the dynamic response factor. These figures clearly demonstrate that the effect of depth ratio is significant for both massless ($B_z = 0$) and massive foundations. As expected, with the increase in the depth ratio, the foundation becomes stiffer, i.e., the dynamic response factor decreases. Another feature of the solutions in Figs. 4.1.10 and 4.1.11 is the negligible dependence on the dynamic response factor for a massive foundation at higher frequencies away from the resonant peak.

Effect of Mass Ratio of the Foundation

To investigate the effect of the mass ratio of the foundation on the dynamic interaction, Fig. 4.1.12A and B presents the dynamic response factor of an embedded foundation with a depth ratio l=0.5 and different values of mass ratio B_z (B_z = 0, 1, 2, and 4) for H/r_0 = 2 and 20, respectively. The results indicate that both the location and the shape of the resonant peaks were quite sensitive to B_z. By increasing the values of B_z, the peak amplitude of the dynamic response factor increases while the resonant frequency shifts to a lower value. In addition, the greater the mass ratio, the narrower and steeper is the corresponding resonant peak curve.

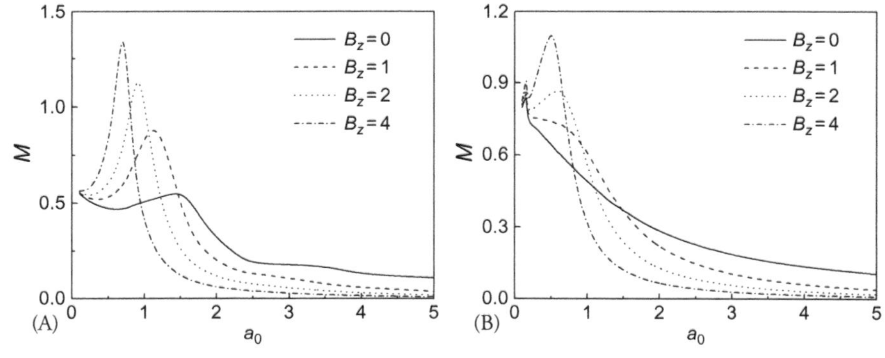

Fig. 4.1.12 Effect of mass ratio on the dynamic response factor for an embedded foundation (l = 0.5). (A) H/r_0 = 2 and (B) H/r_0 = 20

Effect of Internal Friction b^*

It is necessary to study the effect of the poroelastic material parameter b^*, which represents the internal friction due to the relative motion between the solid matrix and the pore fluid, since a salient feature of the poroelastic soil is the generation and dissipation of pore pressure under applied loading. The dynamic response factor of a rigid foundation embedded in a soil layer with different b^* ($b^* = 0.1$, 1, 10, 100, and 1000) is shown in Figs. 4.1.13–4.1.15 for three values of soil-layer thickness ($H/r_0 = 2$, 10, and 20). Solutions are given for $l = 0.5$ and two values of mass ratio, $B_z = 0$ and 2. Note that the parameter b^* is associated with the ratio between the fluid viscosity and the soil permeability. Since the fluid viscosity can be assumed to be identical for most poroelastic mediums at a fixed temperature, any difference in b^* among the five poroelastic materials can be considered as a difference in the soil permeability. A larger value of b^* can be interpreted as a smaller value of soil permeability. Therefore, the material with $b^* = 0.1$ is the most permeable while the material with $b^* = 1000$ is the least permeable among the five poroelastic soil layers.

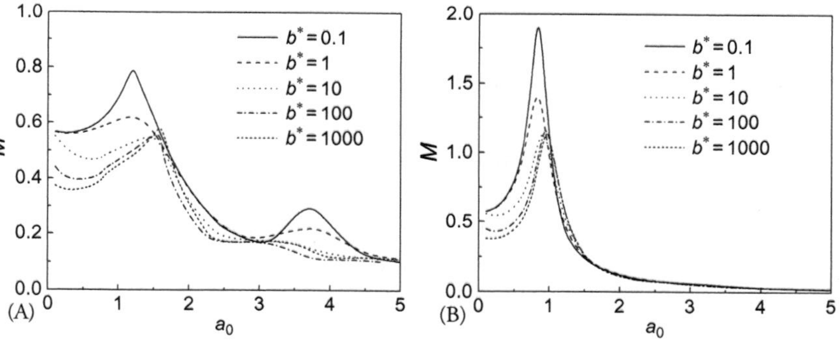

Fig. 4.1.13 Effect of internal friction on the dynamic response factor for an embedded foundation ($l = 0.5$, $H/r_0 = 2$). (A) $B_z = 0$ and (B) $B_z = 2$

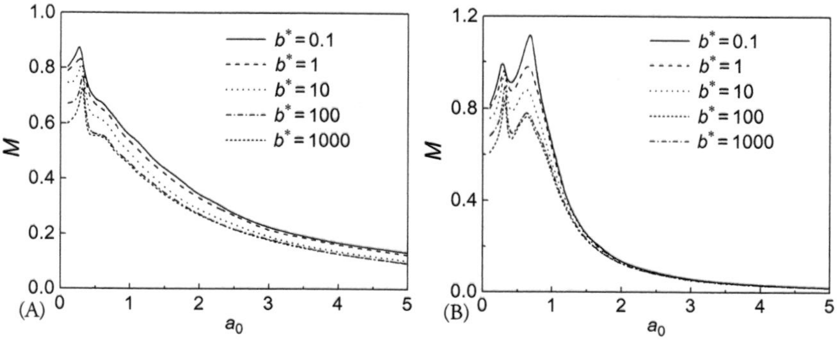

Fig. 4.1.14 Effect of internal friction on the dynamic response factor for an embedded foundation ($l = 0.5$, $H/r_0 = 10$). (A) $B_z = 0$ and (B) $B_z = 2$

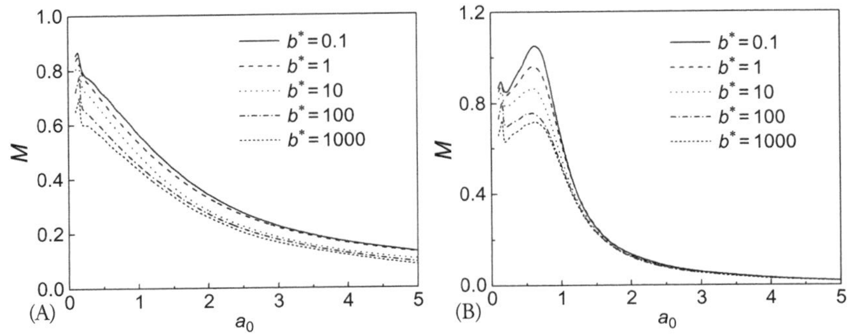

Fig. 4.1.15 Effect of internal friction on the dynamic response factor for an embedded foundation (l = 0.5, H/r_0 = 20). (A) B_z = 0 and (B) B_z = 2

The material property $b*$ was found to have a significant effect on the dynamic response factor. For a shallow soil layer, as shown in Fig. 4.1.13, the dynamic response factor of a massless foundation showed fluctuations with frequency a_0 and the fluctuation was more pronounced when $b*$ was small. As $b*$ increased, the first resonant frequency increased and any increase in $b*$ beyond $b*$ = 100 did not affect the response significantly. This is reasonable since a medium is practically fully impermeable when it is beyond a certain value of permeability. Generally, materials such as sand may have a low value of $b*$, and clays have a larger value of $b*$. A similar behavior was also noted by Rajapaks and Senjuntichai[33] for the case of dynamic response of a multilayered poroelastic medium when subjected to time-harmonic loading applied in the interior of the layered medium. For the soil layer with H/r_0 = 10 and 20, as shown in Figs. 4.1.14 and 4.1.15, the dynamic response factor also showed a notable dependence on $b*$, and the general trend was a decrease in the dynamic response factor, as $b*$ increased over the entire frequency range presented ($0.1 \le a_0 \le 5$). Comparisons between the cases of B_z = 0 and 2 indicate that, for a massive system, $b*$ has little effect on the dynamic response factor at frequencies away from the resonant peak.

Comparisons Between Poroelastic Soil Layer and Elastic Soil Layer

In order to demonstrate the effect of pore fluid, Fig. 4.1.16 presents a comparison of the dynamic response factor between a poroelastic soil layer ($b*$ = 10) and an elastic soil layer for a different depth ratio (l = 0.25, 0.5, and 1.0).

The comparison was conducted for a layer thickness H/r_0 = 20 and a mass ratio B_z = 4. It is clear that the effect of pore fluid was to considerably decrease the dynamic response factor and accordingly contributed to the reduction of vibration response. This phenomenon was particularly pronounced for frequencies close to the resonant value. The results shown in Fig. 4.1.16 clearly indicate that the elastic solutions cannot be used to approximate the vertical dynamic response factor of foundations embedded in a poroelastic soil layer, except at high frequencies away from the resonant frequency.

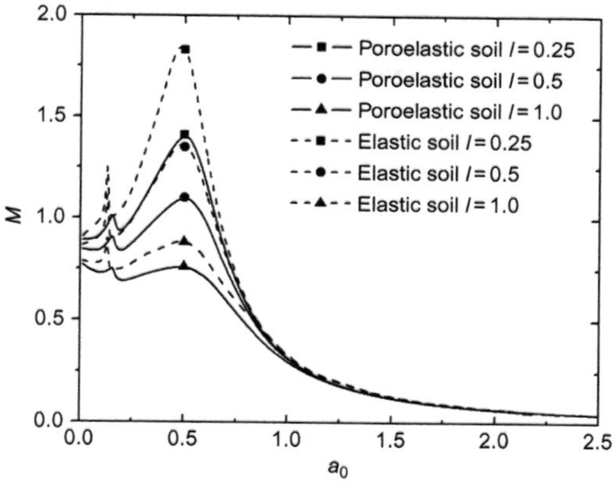

Fig. 4.1.16 Comparison of the dynamic response factor between a poroelastic soil layer and an elastic soil layer ($H/r_0 = 20$, $B_z = 4$)

4.1.5 Conclusions

The vertical dynamic response of a rigid, massive, cylindrical foundation embedded in a poroelastic soil layer based on rigid bedrock was investigated through a simplified method. The foundation was subjected to time-harmonic vertical loading and was perfectly bonded to the surrounding soil in the vertical direction. Biot's theory of poroelastodynamics was used in the analysis. The present numerical solutions are stable and are in good agreement with the well-known solutions obtained from other approaches for both the elastodynamic interaction problem and poroelastodynamic interaction problem. Selected numerical results were presented for vertical dynamic impedance and dynamic response factor.

Numerical results presented in this paper indicate that the frequency of excitation, the thickness of the soil layer, the depth of the embedment, the mass ratio of the foundation and the poroelastic material parameters all have a significant effect on the dynamic interaction between a rigid embedded foundation and a poroelastic soil layer. Both dynamic impedance and the response factor of the foundations embedded in a poroelastic soil layer showed significant fluctuations with the nondimensional frequency of excitation. The shallower the soil layer, the more pronounced was the fluctuation. As the thickness of the soil layer increased, the dynamic response of the soil layer showed a gradual agreement with that of the half-space and the effect of the thickness of the soil layer was found to be almost negligible when $H/r_0 \geq 20$. An increase in the dynamic impedance and a decrease in the dynamic response factor were observed with the increase in the depth ratio. The numerical analysis showed that both the location and the shape of the resonant peak of the dynamic response factor revealed a strong dependence on the mass ratio of the foundation. By increasing the mass ratio, the peak amplitude of the dynamic response factor increased as well, while the resonant frequency decreased. The larger the mass ratio, the narrower and steeper was the corresponding resonant peak curve. In addition, the pore fluid had a significant effect on the dynamic interaction problem. A decrease in the dynamic response factor was noted due to the presence of pore fluid. The simplified yet easy-to-use analytical method presented in this paper can potentially be used to study other dynamic interaction problems involving a rigid embedded foundation and a poroelastic half-space.

Chapter 5
Dynamic Responses of Foundations Under Elastic Waves

This chapter introduces problem-solving procedures for the dynamic interaction between saturated ground and foundations under elastic waves.

5.1 Dynamic Response of a Rigid Foundation on Saturated Soil to Plane Waves

5.1.1 Governing Equations of Poroelastic Soil

The governing equations for poroelastic half-space are given in Eqs. (1.1.24), (1.1.25).

Based on the Helmholtz decomposition, u and w have the following decomposition:

$$\begin{aligned} u &= \text{grad } \varphi + \text{curl } \psi \\ w &= \text{grad } \chi + \text{curl } \gamma \end{aligned} \tag{5.1.1}$$

where φ and ψ are scalar and vector potentials of solid skeleton, respectively, and χ and γ are scalar and vector potentials of pore fluid, respectively.

According to the analysis of Lin et al.[34], by substituting Eq. (5.1.1) into Eqs. (1.1.24), (1.1.25), potentials for the porous medium can be determined.

5.1.2 The Total Wave Fields in Saturated Soil

Incident P Wave

The seismic excitation, which is represented by incident plane fast P waves, as shown in Fig. 5.1.1, can be expressed by the potential

$$\varphi_1^{\text{in}} = a_0 \mathrm{e}^{-\mathrm{i}k_{\text{p1}}(x\sin\theta_{\text{in}} - z\cos\theta_{\text{in}})} \tag{5.1.2}$$

For an incident plane fast P wave, the reflection from the free surface of the poroelastic half-space generates three reflected plane waves: fast and slow P waves, and SV waves. A general representation

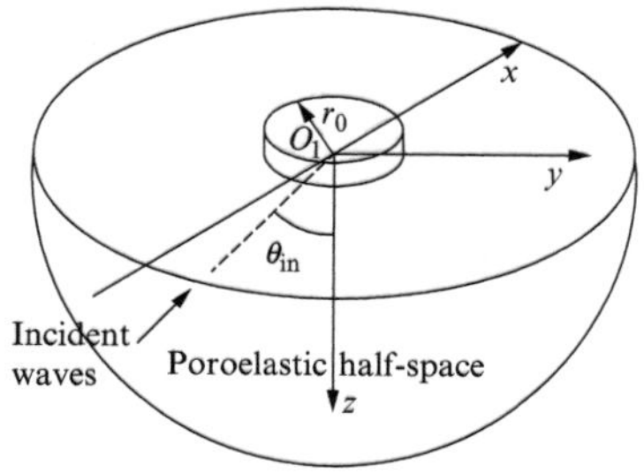

Fig. 5.1. 1 Rigid circular foundation on poroelastic half-space and the incident waves

for the reflected wave is

$$\varphi_1^{re}=a_1 e^{-ik_{p1}(x\sin\theta_{p1}+z\cos\theta_{p1})} \tag{5.1.3}$$

$$\varphi_2^{re}=a_2 e^{-ik_{p2}(x\sin\theta_{p2}+z\cos\theta_{p2})} \tag{5.1.4}$$

$$\psi^{re}=b_1 e^{-ik_s(x\sin\theta_s+z\cos\theta_s)} \tag{5.1.5}$$

where a_0 is the amplitude of the incident wave; θ_{in} is the angle of incidence; a_1, a_2, and b_1 are the amplitudes of the reflected fast P, slow P, and SV waves, respectively; θ_{p1}, θ_{p2}, and θ_s are the angles of the reflected fast P, slow P, and SV waves; moreover, $\theta_{in} = \theta_{p1}$; k_{p1}, k_{p2}, and k_s are the wave numbers of the fast P, slow P, and SV waves in the medium. Expressions for the amplitudes, angles, and wave numbers of the reflected waves can be found in Lin et al.[34]

The potentials can be represented in the coordinate system $O(r, \theta, z)$:

$$\varphi_1^{in}=a_0 e^{ik_{p1z}z-ik_e r\cos\theta} \tag{5.1.6}$$

$$\varphi_1^{re}=a_1 e^{-ik_{p1z}z-ik_e r\cos\theta} \tag{5.1.7}$$

$$\varphi_2^{re}=a_2 e^{-ik_{p2z}z-ik_e r\cos\theta} \tag{5.1.8}$$

$$\psi^{re}=b_1 e^{-ik_{sz}z-ik_e r\cos\theta} \tag{5.1.9}$$

where $k_{p1z} = k_{p1}\cos\theta_{p1}$, $k_{p2z} = k_{p2}\cos\theta_{p2}$, $k_{sz} = k_s\cos\theta_s$, and $k_e = k_{p1}\sin\theta_{p1} = k_{p2}\sin\theta_{p2} = k_s\sin\theta_s$.

In the preceding equations, φ_1^{in}, φ_1^{re}, φ_2^{re}, and ψ^{re} are considered to be free-field waves describing the wave propagation in poroelastic half-space in the absence of foundations. The presence of the rigid foundation, which can be regarded as a secondary wave source, modifies the free-field motion and gives rise to the scattering waves. According to the scattering theory of elastic waves suggested by Pao and Chao[35], the total scattering wave field in soil can be divided into two parts. One corresponds to rigid-body scattering waves generated when the free-field waves are scattered by the fixed rigid foundation on the poroelastic half-space, which is denoted as u^S. And the other one is regarded as radiation scattering waves generated by the vibration of the rigid foundation, which is denoted as u^{R}. Therefore, the total wave fields consist of free-field waves, rigid-body scattering waves and radiation scattering waves.

Incident SV Wave

When the seismic excitation is presented by SV waves, there are two possible critical angles for SV wave incidence, because two P waves coexist in the poroelastic medium. As the incident angle reaches the critical angle, the reflected P wave becomes a surface wave. When the incident angle exceeds the critical angle, the reflected P wave will decrease exponentially with depth. The critical angles can be determined from

$$\begin{aligned}\theta_{cr1}&=\sin^{-1}(k_{p1}/k_{s})\\ \theta_{cr2}&=\sin^{-1}(k_{p2}/k_{s})\end{aligned} \tag{5.1.10}$$

Considering the critical angles, the free-field waves can be expressed as

$$\psi^{in}=b_0 e^{k_{sz}z-ik_e r\cos\theta} \tag{5.1.11}$$

$$\varphi_1^{re}=a_1 e^{-k_{p1z}z-ik_e r\cos\theta} \tag{5.1.12}$$

$$\varphi_2^{re}=a_2 e^{-k_{p2z}z-ik_e r\cos\theta} \tag{5.1.13}$$

$$\psi^{re}=b_1 e^{k_{sz}z-ik_e r\cos\theta} \tag{5.1.14}$$

where $k_e = k_{p1}\sin\theta_{p1} = k_{p2}\sin\theta_{p2} = k_s\sin\theta_s$; $k_{p1z}=\mathrm{i}\sqrt{k_{p1}^2-k_e^2}$ when $\theta_s < \theta_{cr1}$ and $k_{p1z}=-\sqrt{k_e^2-k_{p1}^2}$ when $\theta_s \geq \theta_{cr1}$; $k_{p2z}=\mathrm{i}\sqrt{k_{p2}^2-k_e^2}$ when $\theta_s < \theta_{cr2}$ and $k_{p2z}=-\sqrt{k_e^2-k_{p2}^2}$ when $\theta_s \geq \theta_{cr2}$; $k_{sz} = k_s\cos\theta_s$.

When the incident waves are SV waves, the total wave fields also consist of free-field waves, rigid-body scattering waves and radiation scattering waves here.

Incident Rayleigh Wave

When Rayleigh waves propagate toward the foundation, such potentials can be given by means of

$$\varphi^{in}=(a_1 e^{-k_r\gamma_1 z}+a_2 e^{-k_r\gamma_2 z})\cdot e^{-ik_r r\cos\theta} \tag{5.1.15}$$

$$\psi^{in}=b_1 e^{-k_r\gamma_3 z}\cdot e^{-ik_r r\cos\theta} \tag{5.1.16}$$

where k_r is the wave number of Rayleigh waves, $\gamma_1=1-c^2/v_{p1}^2$, $\gamma_2=1-c^2/v_{p2}^2$, v_{p1} and v_{p2} denote the velocities of the dilatational wave of the first and second kind, respectively; $\gamma_3=1-c^2/v_s^2$, c is the phase velocity of the waves. These parameters have been derived in detail by Cai et al.[36]

5.1.3 The Vertical Vibration of the Foundation

The system under study is shown in Fig. 5.1.1. The soil medium was modeled as a poroelastic half-space with an unsealed boundary, which was fully saturated by the viscous fluid. A rigid circular foundation, with the center of curvature at O_1 and radius r_0, rests on the poroelastic half-space. The foundation was subjected to the action of obliquely incident plane waves. In what follows, the excitation and the response will have harmonic time dependence of the type $e^{i\omega t}$ (ω is the frequency of the motion and $i = \sqrt{-1}$). For brevity, the factor $e^{i\omega t}$ will be dropped from all expressions.

Making use of the axisymmetry of vertical motion, the governing equations and constitutive relations for homogeneous poroelastic soil in the axisymmetry problem is discussed in Section 4.1.1. The nondimensional frequency and material properties are given in Eq. (4.1.10).

The Hankel transform with respect to the radial coordinate and its inverse Hankel transform were employed to solve the problem, which are defined as[27]

$$\tilde{f}^\nu(\varepsilon) = \int_0^\infty rf(r)J_\nu(\varepsilon r)dr \tag{5.1.17}$$

$$\tilde{f}^\nu(r) = \int_0^\infty \varepsilon f(\varepsilon)J_\nu(\varepsilon r)d\varepsilon \tag{5.1.18}$$

where $J_\nu(\varepsilon r)$ is the νth-order Bessel function of the first kind.

Substituting the dimensionless parameters into Eqs. (4.1.1)–(4.1.9) and applying the Hankel transforms to them, the following general solutions can be obtained based on the procedure proposed by Cai and Hu[37]:

$$\tilde{e}^0(\varepsilon, z^*) = A_1 e^{-q_1 z^*} + A_2 e^{-q_2 z^*} \tag{5.1.19}$$

$$\tilde{p}_f^0(\varepsilon, z^*) = c_{21}A_1 e^{-q_1 z^*} + c_{22}A_2 e^{-q_2 z^*} \tag{5.1.20}$$

$$\tilde{u}_r^1(\varepsilon, z^*) = c_{31}A_1 e^{-q_1 z^*} + c_{32}A_2 e^{-q_2 z^*} + A_3 e^{-q_3 z^*} \tag{5.1.21}$$

$$\tilde{u}_z^0(\varepsilon, z^*) = c_{41}A_1 e^{-q_1 z^*} + c_{42}A_2 e^{-q_2 z^*} + c_{43}A_3 e^{-q_3 z^*} \tag{5.1.22}$$

$$\tilde{\sigma}_{zz}^0(\varepsilon, z^*) = (\lambda - 2q_1 c_{41})A_1 e^{-q_1 z^*} + (\lambda - 2q_2 c_{42})A_2 e^{-q_2 z^*} - 2q_3 c_{43} e^{-q_3 z^*} \tag{5.1.23}$$

$$\tilde{\sigma}_{rz}^1(\varepsilon, z) = -(q_1 c_{31} + \varepsilon c_{41})A_1 e^{-q_1 z^*} - (q_2 c_{32} + \varepsilon c_{42})A_2 e^{-q_2 z^*} - (q_3 + \varepsilon c_{43})e^{-q_3 z^*} \tag{5.1.24}$$

where A_1, A_2, and A_3 are arbitrary functions to be determined from the boundary conditions. Expressions for q_1, q_2, q_3, c_{21}, c_{22}, c_{31}, c_{32}, c_{41}, c_{41}, and c_{43} are given in Appendix E.

At $z = 0$, the perfect bond between the soil and the foundation can be stated as follows:

$$\sigma_{rz}(r^*, 0) = 0 \quad (0 \le r^* \le \infty) \tag{5.1.25}$$

$$u_z(r^*, 0) = A_{T0} \quad (0 \le r^* \le 1) \tag{5.1.26}$$

where A_{T0} denotes the nondimensional vertical displacement of the rigid foundation.

Two types of drainage conditions were considered at the surface of the poroelastic half-space. An unsealed half-space boundary implies

$$p_f(r^*, 0)=0 \quad (0<r^*<\infty) \tag{5.1.27}$$

$$\sigma_{zz}(r^*, 0)=0 \quad (1<r^*<\infty) \tag{5.1.28}$$

and for a sealed boundary, it is

$$\sigma_{zz}(r^*, 0)+p_f(r^*, 0)=0 \quad (1<r^*<\infty) \tag{5.1.29}$$

$$\frac{p_f(r^*, z^*)}{z^*}=0 \quad (0<r^*<\infty) \tag{5.1.30}$$

For the unsealed boundary, following the definition of rigid-body scattering waves and radiation scattering waves, Eqs. (5.1.26), (5.1.28) can be rewritten in terms of u^R and u^S as

$$u_z^R(r^*, 0)=A_{T0} \quad (0\le r^*\le 1) \tag{5.1.31}$$

$$\sigma_{zz}^R(r^*, 0)=0 \quad (1<r^*<\infty) \tag{5.1.32}$$

$$u_z^S(r^*, 0)=-u_z^{\text{in+re}}(r^*, 0) \quad (0\le r^*\le 1) \tag{5.1.33}$$

$$\sigma_{zz}^S(r^*, 0)=0 \quad (1<r^*<\infty) \tag{5.1.34}$$

and for the sealed boundary

$$u_z^R(r^*, 0)=A_{T0} \quad (0\le r^*\le 1) \tag{5.1.35}$$

$$\sigma_{zz}^R(r^*, 0)+p_f^R(r^*, 0)=0 \quad (1<r^*<\infty) \tag{5.1.36}$$

$$u_z^S(r^*, 0)=-u_z^{\text{in+re}}(r^*, 0) \quad (0\le r^*\le 1) \tag{5.1.37}$$

$$\sigma_{zz}^S(r^*, 0)+p_f^S(r^*, 0)=0 \quad (1<r^*<\infty) \tag{5.1.38}$$

where $u_z^{\text{in+re}}$, u_z^R, and u_z^S are nondimensional vertical displacement components of the free-field wave, radiation scattering wave, and rigid-body scattering wave, respectively.

When the incident waves are P waves, Huang et al.[38] have given the vertical displacement of the free-field wave at the saturated half-space surface:

$$u_z^{\text{in+re}}(r^*, 0)=\mathrm{i}k_{p1z}(a_0-a_1)J_0(k_e r^*)-\mathrm{i}k_{p2z}a_2J_0(k_e r^*)+\frac{b_1k_e^2}{k_s}J_0(k_e r^*) \tag{5.1.39}$$

Applying the inverse Hankel transform to Eqs. (5.1.19)–(5.1.24), together with Eqs. (5.1.31)–(5.1.38), the following two sets of standard dual integral equations can be derived:

$$\begin{aligned}&\int_0^\infty \varepsilon^{-1}(1+H(\varepsilon))B_{\mathrm{R}}(\varepsilon)J_0(\varepsilon r^*)\mathrm{d}\varepsilon = A_{\mathrm{T0}}/\eta \qquad (0\le r^*\le 1)\\ &\int_0^\infty B_{\mathrm{R}}(\varepsilon)J_0(\varepsilon r^*)\mathrm{d}\varepsilon = 0 \qquad (1<r^*<\infty)\end{aligned} \tag{5.1.40}$$

$$\begin{aligned}&\int_0^\infty \varepsilon^{-1}(1+H(\varepsilon))B_{\mathrm{s}}(\varepsilon)J_0(\varepsilon r^*)\mathrm{d}\varepsilon = -u_z^{\mathrm{in+re}}/\eta \qquad (0\le r^*\le 1)\\ &\int_0^\infty B_{\mathrm{s}}(\varepsilon)J_0(\varepsilon r^*)\mathrm{d}\varepsilon = 0 \qquad (1<r^*<\infty)\end{aligned} \tag{5.1.41}$$

where $B_{\mathrm{R}}(\varepsilon)=\varepsilon\tilde{\sigma}^0_{zz\mathrm{R}}(\varepsilon,0)$, $B_{\mathrm{s}}(\varepsilon)=\varepsilon\tilde{\sigma}^0_{zz\mathrm{s}}(\varepsilon,0)$, $H(\varepsilon)=\dfrac{\varepsilon f(\varepsilon)}{\eta}-1$, $\eta=\lim\limits_{\varepsilon\to\infty}\varepsilon f(\varepsilon)$, expressions for $f(\varepsilon)$ are given in Appendix E.

According to the methods suggested by Nobel[39], Eqs. (5.1.40), (5.1.41) can be reduced to Fredholm integral equations of the second kind:

$$B_{\mathrm{R}}(\varepsilon)=\frac{2\varepsilon A_{\mathrm{T0}}}{\pi\eta}\int_0^1 \theta_{\mathrm{R}}(x)\cos(\varepsilon x)\mathrm{d}x \tag{5.1.42}$$

Substituting Eq. (5.1.42) into Eq. (5.1.40), the following integral equation of the Fredholm type can be obtained:

$$\theta_{\mathrm{R}}(x)+\frac{1}{\pi}\int_0^1 M(x,y)\theta_{\mathrm{R}}(y)\mathrm{d}y = 1 \tag{5.1.43}$$

where the kernel function $M(x, y)$ takes the form

$$M(x,y)=2\int_0^\infty H(\varepsilon)\cos(\varepsilon x)\cos(\varepsilon y)\mathrm{d}\varepsilon \tag{5.1.44}$$

Employing the following integral representation

$$B_{\mathrm{s}}(\varepsilon)=\frac{2\varepsilon}{\pi}\int_0^1 \theta_{\mathrm{s}}(x)\cos(\varepsilon x)\mathrm{d}x \tag{5.1.45}$$

Substituting Eqs. (5.1.39), (5.1.45) into Eq. (5.1.41), the following integral equation of the Fredholm type can be obtained:

$$\theta_{\mathrm{s}}(x)+\frac{1}{\pi}\int_0^1 M(x,y)\theta_{\mathrm{s}}(y)\mathrm{d}y = \left(-\mathrm{i}(a_0-a_1)k_{\mathrm{p1z}}+\mathrm{i}a_2k_{\mathrm{p2z}}-\frac{b_1k_{\mathrm{e}}^2}{k_{\mathrm{s}}}\right)\cos(k_{\mathrm{e}}r^*) \tag{5.1.46}$$

The kernel function $M(x, y)$ is given by Eq. (5.1.44).

Eqs. (5.1.43), (5.1.46) can be discretized into a series of algebraic equations and can be solved numerically. Combining with Eqs. (5.1.42), (5.1.45), the total force acting at the base T of the foundation can be obtained:

$$\frac{T}{\mu r_0^2}=\frac{4A_{\mathrm{T0}}}{\eta}\int_0^1\theta_{\mathrm{R}}(x)\mathrm{d}x+\frac{4}{\eta}\int_0^1\theta_{\mathrm{S}}(x)\mathrm{d}x \tag{5.1.47}$$

where $T^*=T/\mu r_0^2$ is the nondimensional form of T.

Eq. (5.1.47) indicates that the force exerted on the foundation may be thought of as being composed of two parts: the first part corresponds to the force required to move the foundation by displacement U_0 when there is no free-field motion, and the second part corresponds to the force to hold the foundation in place when it is subjected to the action of the free-field waves.

We introduce the definitions

$$K_{\mathrm{TT}}=\frac{4}{\eta}\int_0^1\theta_{\mathrm{R}}(t)\mathrm{d}t \tag{5.1.48}$$

and

$$U^*=-\frac{\frac{4}{\eta}\int_0^1\theta_{\mathrm{S}}(t)\mathrm{d}t}{K_{\mathrm{TT}}} \tag{5.1.49}$$

Eq. (5.1.47) can be rewritten in the form

$$T^*=K_{\mathrm{TT}}\,(A_{\mathrm{T0}}-U^*) \tag{5.1.50}$$

where K_{TT} is the vertical dynamic impedance for the foundation. U^* is called input vertical motion here and corresponds to the vertical vibrations of the rigid foundation under the action of the seismic excitation when no external forces are acting on the foundation ($T^*=0$). Eq. (5.1.50) establishes the force and vertical vibration relationship for the rigid foundation subjected to the action force and the seismic excitation.

By now, only the foundation vertical displacement U_0 remains unknown. It can be determined from the dynamic equilibrium condition for the foundation. Combining with Eq. (5.1.47), the dynamic equilibrium implies

$$-\omega_0^2 m^* A_{\mathrm{T0}}=\frac{4A_{\mathrm{T0}}}{\eta}\int_0^1\theta_{\mathrm{R}}(x)\mathrm{d}x+\frac{4}{\eta}\int_0^1\theta_{\mathrm{S}}(x)\mathrm{d}x \tag{5.1.51}$$

where m^* is the dimensionless mass of the foundation, $m^*=m/\rho r_0^3$.

According to the vibration analysis method given by Richart[40], the vertical displacement amplitude of the foundation excited by incident P waves can be written as

$$A_{\mathrm{T0}}=\frac{\left|\frac{4}{\eta}\int_0^1\theta_{\mathrm{S}}(t)\mathrm{d}t\right|}{\sqrt{(k_{\mathrm{t}}-m^*\omega_0^2)^2+(c_{\mathrm{t}}\omega_0)^2}} \tag{5.1.52}$$

where $k_{\mathrm{t}}=-\mathrm{Re}(K_{\mathrm{TT}})$, $c_{\mathrm{t}}=-\mathrm{Im}(K_{\mathrm{TT}})/\omega_0$.

Likewise, the vertical vibrations of the foundation excited by incident SV waves can be obtained by employing the same procedure. The only difference is that Eqs. (5.1.39), (5.1.46) have the following forms:

$$u_z^{\text{in}+\text{re}}(r^*,0) = -k_{\text{p1z}}a_1J_0(k_{\text{e}}r^*) - k_{\text{p2z}}a_2J_0(k_{\text{e}}r^*) + \frac{(b_0+b_1)k_{\text{e}}^2}{k_{\text{s}}}J_0(k_{\text{e}}r^*) \tag{5.1.53}$$

$$\theta_{\text{s}}(x) + \frac{1}{\pi}\int_0^1 M(x,y)\theta_{\text{s}}(y)\text{d}y = \left(a_1k_{\text{p1z}} + a_2k_{\text{p2z}} - \frac{(b_0+b_1)k_{\text{e}}^2}{k_{\text{s}}}\right)\cos(k_{\text{e}}r^*) \tag{5.1.54}$$

For brevity, while without loss of generality, the gravity force of the foundation is neglected in the preceding derivations.

5.1.4 The Rocking Vibration of the Foundation

The model considered here is demonstrated in Fig. 5.1.2. A rigid circular foundation, with the center of curvature at O_1 and radius r_0, rests on the poroelastic half-space. The foundation was subjected to the action of P and SV waves.

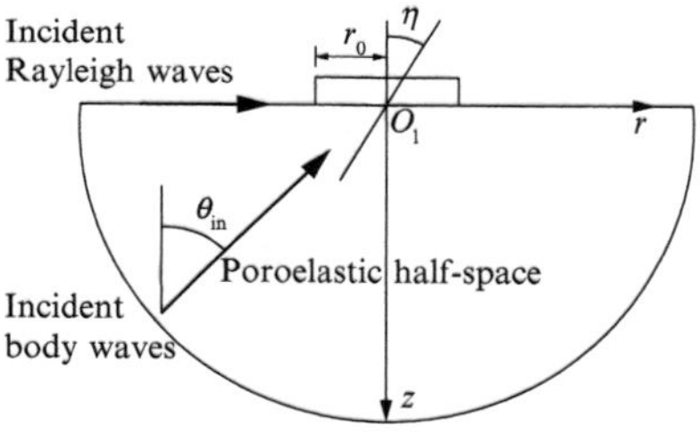

Fig. 5.1.2 Rigid foundation on poroelastic half-space and incident waves

For a rocking vibration, the governing equations of the poroelastic soil in the cylindrical coordinate system $O(r, \theta, z)$ are given in Eq. (1.1.28).

The constitutive relations for a homogeneous poroelastic material are given by Eq. (1.1.21).

It was assumed that the foundation was perfectly bonded to the soil along the side of the foundation. The contact surface between the soil and the foundation was smooth and fully permeable. For the problem of rocking vibration, boundary conditions of the soil are given as follows:

$$\tau_{zr}(r, \theta, 0)=0 \quad (0 \le r < \infty) \tag{5.1.55}$$

$$\tau_{zr}(r, \theta, 0)=0 \quad (0 \le r < \infty) \tag{5.1.56}$$

$$\tau_{z\theta}(r, \theta, 0)=0 \quad (0 \le r < \infty) \tag{5.1.57}$$

$$p_{\text{f}}(r, \theta, 0)=0 \quad (0 \le r < \infty) \tag{5.1.58}$$

$$\sigma_{zz}(r,\theta,0)=0 \quad (r_0<r<\infty) \tag{5.1.59}$$

$$u_z(r,\theta,0)=\eta r\cos\theta \quad (0\le r\le r_0) \tag{5.1.60}$$

where η denotes the rocking angle amplitude of the rigid foundation.

The displacements, stress, strain, and pore pressure can be expressed as follows:

$$\begin{aligned}&\left\{\begin{array}{l}u_r(r,\theta,z)\\u_z(r,\theta,z)\\\tau_{rz}(r,\theta,z)\\\sigma_{zz}(r,\theta,z)\\e(r,\theta,z)\\p_f(r,\theta,z)\end{array}\right\}=\sum_{n=0}^{\infty}\left\{\begin{array}{l}u_{rn}(r,z)\\u_{zn}(r,z)\\\tau_{rzn}(r,z)\\\sigma_{zn}(r,z)\\e_n(r,z)\\p_{fn}(r,z)\end{array}\right\}\cos(n\theta)\\&\left\{\begin{array}{l}u_\theta(r,\theta,z)\\\tau_{\theta z}(r,\theta,z)\end{array}\right\}=\sum_{n=0}^{\infty}\left\{\begin{array}{l}u_{\theta n}(r,z)\\\tau_{\theta zn}(r,z)\end{array}\right\}\sin(n\theta)\end{aligned} \tag{5.1.61}$$

Introducing the dimensionless parameters and applying Hankel transforms to Eqs. (1.1.28), (1.1.21), the general solutions can be obtained as the preceding procedure:

$$\tilde{p}_{fn}^{n}(\varepsilon,z^*)=B_{1n}e^{-g_1z^*}+B_{2n}e^{-g_2z^*} \tag{5.1.62}$$

$$\tilde{e}_n^{n}(\varepsilon,z^*)=l_{21}B_{1n}e^{-g_1z^*}+l_{22}B_{2n}e^{-g_2z^*} \tag{5.1.63}$$

$$\tilde{u}_{zn}^{n}(\varepsilon,z^*)=l_{31}g_1B_{1n}e^{-g_1z^*}+l_{32}g_2B_{2n}e^{-g_2z^*}+B_{3n}e^{-g_3z^*} \tag{5.1.64}$$

$$\tilde{u}_{rn}^{n-1}(\varepsilon,z^*)-\tilde{u}_{\theta n}^{n-1}(\varepsilon,z^*)=-l_{31}\varepsilon B_{1n}e^{-g_1z^*}-l_{32}\varepsilon B_{2n}e^{-g_2z^*}+B_{4n}e^{-g_3z^*} \tag{5.1.65}$$

$$\tilde{\tau}_{zrn}^{n-1}(\varepsilon,z^*)-\tilde{\tau}_{z\theta n}^{n-1}(\varepsilon,z^*)=2l_{31}g_1\varepsilon B_{1n}e^{-g_1z^*}+2l_{32}g_2\varepsilon B_{2n}e^{-g_2z^*}+(\varepsilon B_{3n}-g_3B_{4n})e^{-g_3z^*} \tag{5.1.66}$$

$$\tilde{\tau}_{zrn}^{n+1}(\varepsilon,z^*)+\tilde{\tau}_{z\theta n}^{n+1}(\varepsilon,z^*)=l_{41}B_{1n}e^{-g_1z^*}+l_{42}B_{2n}e^{-g_2z^*}-\left[\left(\frac{2}{\varepsilon}g_3^2+\varepsilon\right)B_{3n}+g_3B_{4n}\right]e^{-g_3z^*} \tag{5.1.67}$$

$$\tilde{\sigma}_{zzn}^{n}(\varepsilon,z^*)=l_{51}B_{1n}e^{-g_1z^*}+l_{52}B_{2n}e^{-g_2z^*}-2g_3B_{3n}e^{-g_3z^*} \tag{5.1.68}$$

where B_{1n}, B_{2n}, B_{3n}, and B_{4n} are arbitrary functions to be determined from the boundary conditions. Expressions for g_1, g_2, g_3, l_{21}, l_{22}, l_{31}, l_{32}, l_{41}, l_{42}, l_{51}, and l_{52} are given in Appendix F.

By introducing the division of the wave fields in poroelastic half-space, Eqs. (5.1.59)–(5.1.61) can be rewritten in the following terms:

$$\sum_{n=0}^{\infty}\sigma_{zzn}^{\mathrm{R}}(r^*,0)\cos n\theta=0 \quad (1<r^*<\infty) \tag{5.1.69}$$

$$\sum_{n=0}^{\infty}u_{zn}^{\mathrm{R}}(r^*,0)\cos n\theta=\eta r\cos\theta \quad (0\le r^*\le 1) \tag{5.1.70}$$

$$\sum_{n=0}^{\infty}\sigma_{zzn}^{\mathrm{S}}(r^*,0)\cos n\theta=0 \quad (1<r^*<\infty) \tag{5.1.71}$$

$$\sum_{n=0}^{\infty}u_{zn}^{\mathrm{S}}(r^*,0)\cos n\theta=-u_z^{\mathrm{in+re}}(r^*,\theta,0) \quad (0\le r^*\le 1) \tag{5.1.72}$$

where u_z^{R}, u_z^{S}, and $u_z^{\mathrm{in+re}}$ are dimensionless displacements caused by rigid-body scattering waves, radiation scattering waves, and free-field waves, respectively.

When the foundation is excited by P waves, the dimensionless displacement of the free-field waves $u_z^{\mathrm{in+re}}$ can be expressed as

$$u_z^{\mathrm{in+re}}(r^*,\theta,0)=\sum_{n=0}^{\infty}\left(k_{\mathrm{p1z}}(a_0-a_1)-a_2k_{\mathrm{p2z}}+\frac{bk_{\mathrm{sx}}^2}{k_{\mathrm{s}}}\right)(-\mathrm{i})^n\kappa_nJ_n(k_{\mathrm{e}}r^*)\cos n\theta \tag{5.1.73}$$

where κ_n=1 when n=0; κ_n=2, when $n\neq0$.

From Eqs. (5.1.62) to (5.1.73), together with the linear independence of trigonometric functions, the following dual integral equations can be derived:

$$\begin{aligned}&\int_0^{\infty}\varepsilon^{-1}(1+H(\varepsilon))B_{\mathrm{R}}(\varepsilon)J_1(\varepsilon r^*)\mathrm{d}\varepsilon=\eta/L \quad (0\le r^*\le 1)\\&\int_0^{\infty}B_{\mathrm{R}}(\varepsilon)J_1(\varepsilon r^*)\mathrm{d}\varepsilon=0 \quad (1<r^*<\infty)\end{aligned} \tag{5.1.74}$$

$$\begin{aligned}&\int_0^{\infty}\varepsilon^{-1}(1+H(\varepsilon))B_{\mathrm{s}}(\varepsilon)J_1(\varepsilon r^*)\mathrm{d}\varepsilon=-u_{z1}^{\mathrm{in+re}}/L \quad (0\le r^*\le 1)\\&\int_0^{\infty}B_{\mathrm{s}}(\varepsilon)J_1(\varepsilon r^*)\mathrm{d}\varepsilon=0 \quad (1<r^*<\infty)\end{aligned} \tag{5.1.75}$$

where $u_{z1}^{\mathrm{in+re}}=-2\mathrm{i}\left(k_{\mathrm{p1z}}(a_0-a_1)-a_2k_{\mathrm{p2z}}+\frac{bk_{\mathrm{sx}}^2}{k_{\mathrm{s}}}\right)J_1(k_{\mathrm{e}}\bar{r})$

$$f(\varepsilon)=\frac{-g_1l_{21}+g_2l_{22}+g_1(g_3^2-g_1^2)l_{31}+g_2(g_2^2-g_3^2)l_{32}}{2g_1g_3[l_{21}+(\varepsilon^2+g_1^2)l_{31}]-2g_2g_3[l_{22}+(\varepsilon^2+g_2^2)l_{32}]+(\varepsilon^2+g_3^2)l_{51}-(\varepsilon^2+g_3^2)l_{52}}$$

$$H(\varepsilon)=\frac{\varepsilon f(\varepsilon)}{L}-1,\ L=\lim_{\varepsilon\to\infty}\varepsilon f(\varepsilon)$$

Based on the procedure suggested by Noble[39], Eqs. (5.1.74), (5.1.75) can be reduced to Fredholm integral equations of the second kind:

$$B_{\mathrm{R}}(\varepsilon)=\frac{4\varepsilon\eta}{\pi L}\int_0^1\theta_{\mathrm{R}}(x)\sin(\varepsilon x)\mathrm{d}x$$

$$\theta_{\mathrm{R}}(x)+\frac{1}{\pi}\int_0^1 K(x,y)\theta_{\mathrm{R}}(y)\mathrm{d}y=1 \tag{5.1.76}$$

$$B_{\mathrm{S}}(\varepsilon)=\frac{4\varepsilon}{\pi L}\int_0^1\theta_{\mathrm{S}}(x)\sin(\varepsilon x)\mathrm{d}x$$

$$\theta_{\mathrm{s}}(x)+\frac{1}{\pi}\int_0^1 K(x,y)\theta_{\mathrm{s}}(y)\mathrm{d}y=\left(-(a_0-a_1)k_{\mathrm{p1z}}+a_2k_{\mathrm{p2z}}+\frac{ibk_{\mathrm{sx}}^2}{k_{\mathrm{s}}}\right)\sin(k_{\mathrm{e}}r^*) \tag{5.1.77}$$

where the kernel function $K(x, y)$ takes the form

$$K(x,y)=2\int_0^{\infty}H(\varepsilon)\sin(\varepsilon x)\sin(\varepsilon y)\mathrm{d}\varepsilon \tag{5.1.78}$$

The total moment acting on the foundation base can be obtained as follows:

$$\frac{T}{\mu r_0^3}=\frac{8\eta}{L}\int_0^1 x\theta_{\mathrm{R}}(x)\mathrm{d}x+\frac{8}{L}\int_0^1 x\theta_{\mathrm{S}}(x)\mathrm{d}x \tag{5.1.79}$$

where $T^*=T/\mu r_0^3$ is the dimensionless form of T.

We introduce the definition

$$\eta_0=-\frac{\int_0^1 x\theta_{\mathrm{S}}(x)\mathrm{d}x}{\int_0^1 x\theta_{\mathrm{R}}(x)\mathrm{d}x} \tag{5.1.80}$$

where η_0 is called input rocking motion and corresponds to the rocking vibrations of the rigid foundation under the action of the seismic excitation when no external forces are acting on the foundation.

Introducing the dynamic equilibrium condition for the foundation and combining with Eq. (5.1.79), the following equation can be obtained:

$$-\omega_0^2 m^*\eta=\frac{8\eta}{L}\int_0^1 x\theta_{\mathrm{R}}(x)\mathrm{d}x+\frac{8}{L}\int_0^1 x\theta_{\mathrm{S}}(x)\mathrm{d}x \tag{5.1.81}$$

where m^* is the dimensionless mass of the foundation, $m^*=m/\rho r_0^5$. According to the vibration analysis method[40] the rocking displacement amplitude of the foundation excited by incident P waves can be written as

$$\eta=\frac{\left|\frac{8}{L}\int_0^1 x\theta_{\mathrm{S}}(x)\mathrm{d}x\right|}{\sqrt{(k_{\mathrm{t}}-m^*\omega_0^2)^2+(c_{\mathrm{t}}\omega_0)^2}} \tag{5.1.82}$$

where $k_{\mathrm{t}}=-\mathrm{Re}\left(\frac{8}{L}\int_0^1 x\theta_{\mathrm{R}}(x)\mathrm{d}x\right)$, $c_{\mathrm{t}}=-\mathrm{Im}\left(\frac{8}{L}\int_0^1 x\theta_{\mathrm{R}}(x)\mathrm{d}x\right)/\omega_0$.

Employing the same procedure, the rocking displacement amplitude of the foundation excited by incident SV waves and Rayleigh waves can be obtained accordingly. The difference is that Eq. (5.1.77) takes the following form for incident SV waves:

$$B_s(\varepsilon) = \frac{4\varepsilon}{\pi L}\int_0^1 \theta_R(x)\sin(\varepsilon x)dx$$

$$\theta_s(x) + \frac{1}{\pi}\int_0^1 K(x,y)\theta_s(y)dy = \left(a_1 k_{p1z} + a_2 k_{p2z} + \frac{i(b_0+b_1)k_{sx}^2}{k_s}\right)\sin(k_e r^*) \tag{5.1.83}$$

and for Rayleigh waves

$$B_s(\varepsilon) = \frac{4\varepsilon}{\pi L}\int_0^1 \theta_R(x)\sin(\varepsilon x)dx$$

$$\theta_s(x) + \frac{1}{\pi}\int_0^1 K(x,y)\theta_s(y)dy = \left(-a_1\gamma_1 k_r - a_2\gamma_2 k_r + b_1 k_r^2\right)\sin(k_e r^*) \tag{5.1.84}$$

5.1.5 Numerical Results and Discussion

In this section, the influences of the key parameters on the vertical vibrations of the foundation are discussed. The poroelastic parameters are set to $\alpha = 1$, λ^*=1.004, M^*=246.78, ρ^*=0.45, and n=0.37, which are the same as those given in Cai and Hu[37].

The effect of excitation frequency on the displacement is shown in Figs. 5.1.3 and 5.1.4 for different foundation masses, in which incident angle θ=30°, and internal friction b^*=10. Regardless of the excitations of P waves, SV waves, and Rayleigh waves, for a given value of foundation mass m^*, the displacement increases with increasing excitation frequency ω_0 until it reaches a peak value, and decays quickly after that. This phenomenon is referred to as resonance, and the frequency corresponding to the peak value of the displacement is defined as resonant frequency. It is also observed in both Figs. 5.1.3 and 5.1.4 that the curve is sensitive to the foundation mass. When the excitation frequency is less than the resonant frequency, the increases in foundation mass lead to increases in the displacement amplitude. However, when the excitation frequency is larger than the resonant frequency, the larger the foundation mass, the less significant the vibration of the foundation. In addition, the resonance frequency decreases slightly as the foundation mass increases.

The dynamic responses of elastic and poroelastic soil excited by P waves, SV waves, and Rayleigh waves are presented in Figs. 5.1.5 and 5.1.6, respectively. The incident angle θ_{in}=45°, and the dimensionless mass of the foundation m^*=10. As shown in Figs. 5.1.5 and 5.1.6, the displacements of poroelastic half-space are smaller than those of elastic half-space due to the existence of the pore water in the soil medium. In particular, the resonant amplitude of poroelastic half-space is approximately half that of elastic half-space. In the case of SV wave excitations, as shown in Figs. 5.1.5B and 5.1.6B, a similar phenomenon can also be observed, but the differences between the elastic and poroelastic soil are smaller. The reason is that the propagation of SV waves is independent of pore water, which depends on the solid skeleton only. These results indicate that the elastic solutions can hardly be used to evaluate the performance of the foundation subjected to incident waves.

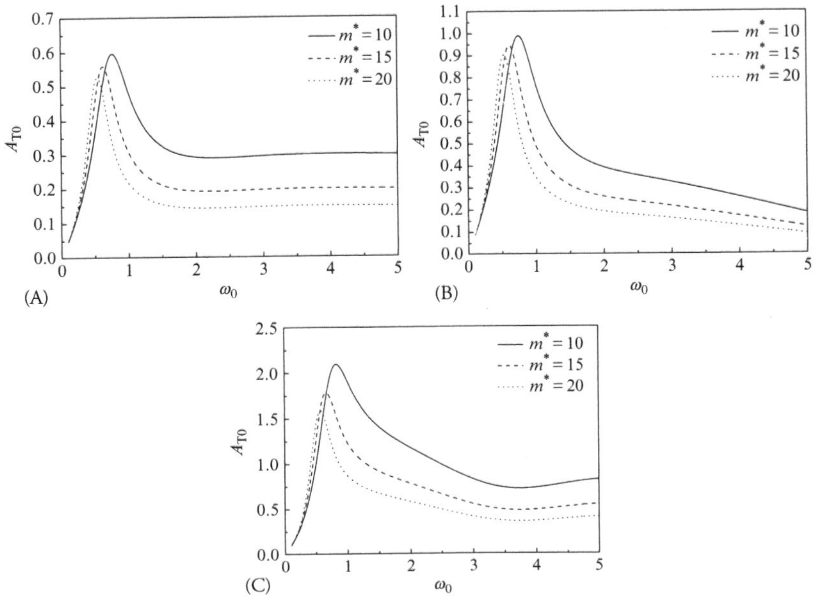

Fig. 5.1.3 Vertical displacement varied with ω_0 under different m^*. (A) excited by P waves, (B) excited by SV waves, and (C) excited by Rayleigh waves

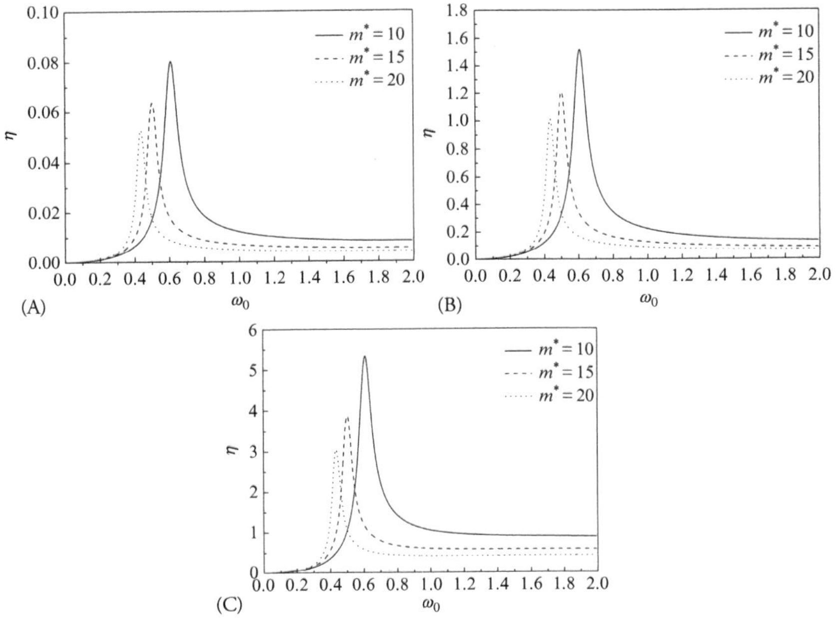

Fig. 5.1.4 Rocking displacement varied with ω_0 under different m^*. (A) excited by P waves, (B) excited by SV waves, and (C) excited by Rayleigh waves

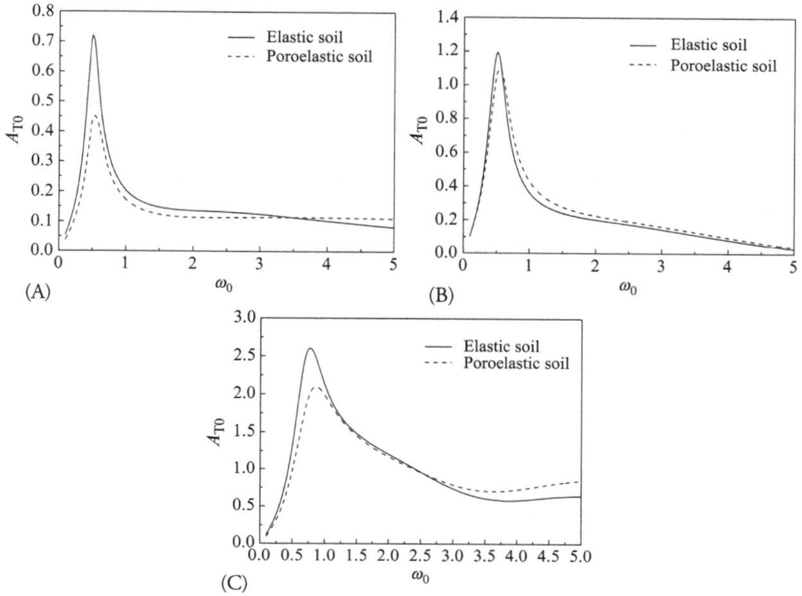

Fig. 5.1.5 Comparison between elastic and poroelastic soil for vertical vibration. (A) excited by P waves, (B) excited by SV waves, and (C) excited by Rayleigh waves

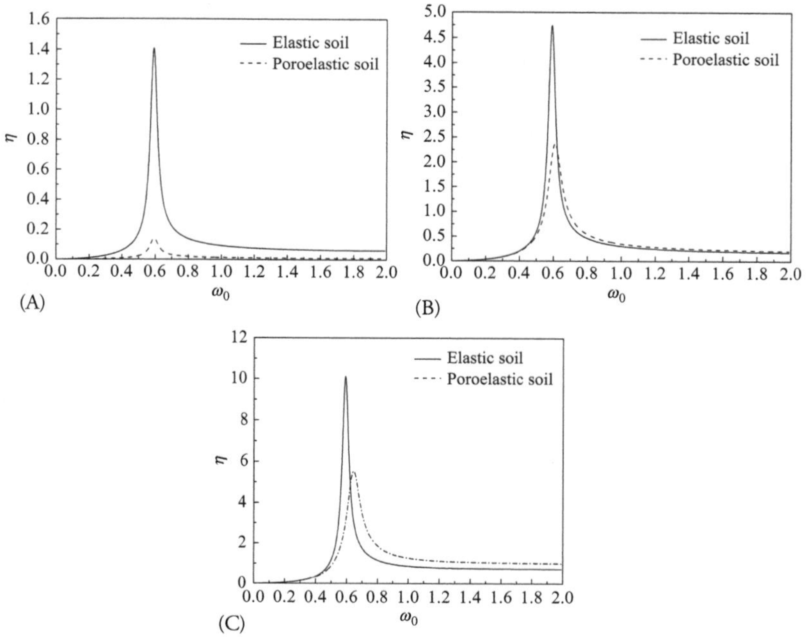

Fig. 5.1.6 Comparison between elastic and poroelastic soil for rocking vibration. (A) excited by P waves, (B) excited by SV waves, and (C) excited by Rayleigh waves

Figs. 5.1.7 and 5.1.8 show the dependence of vertical and rocking displacements on boundary drainage. The internal friction $b^* = 10$; dimensionless mass of the foundation $m^* = 10$. Compared to a sealed boundary, an unsealed boundary contributes to a more significant vertical response. This can be explained by the fact that there is more significant soil-water interaction for the sealed boundary, which gives rise to the energy dissipation.

The internal friction resulting from the relative motion between the solid skeleton and pore fluid is inversely proportional to the intrinsic permeability of the soil medium. The displacement of the foundation subjected to P waves, SV waves, and Rayleigh waves under different internal frictions are shown in Figs. 5.1.9 and 5.1.10, respectively. The incident angle $\theta_{in} = 45°$, and dimensionless mass of the foundation $m^* = 10$. An inspection of Figs. 5.1.9 and 5.1.10 shows that the resonant amplitude decreases as does the internal friction b^*, which is more apparent in rocking vibration. When the excitation frequency is larger than the resonant frequency, the soil with higher intrinsic permeability has a larger vibration amplitude.

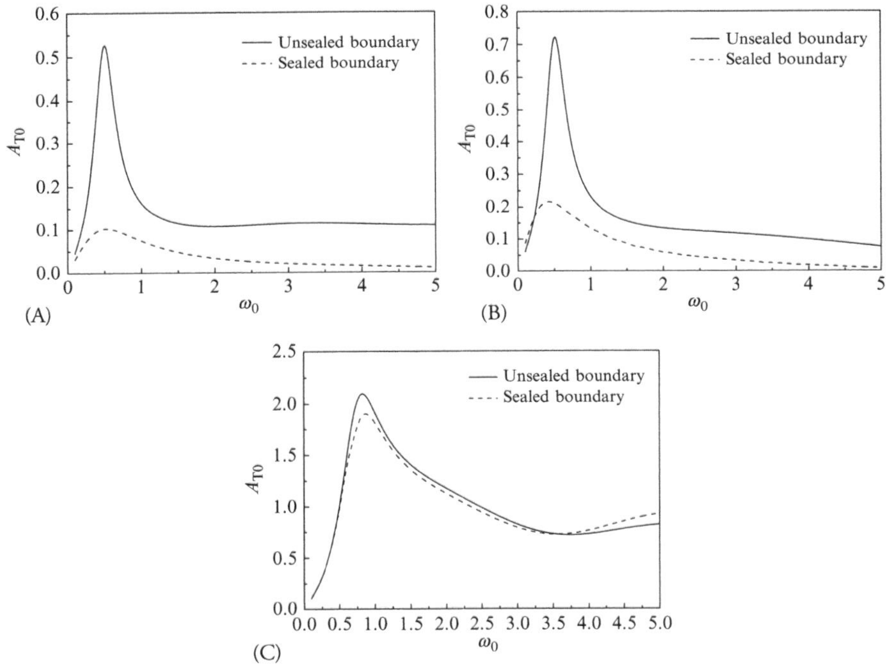

Fig. 5.1.7 Vertical displacement varied with ω_0 under different drainage conditions. (A) excited by P waves, (B) excited by SV waves, and (C) excited by Rayleigh waves

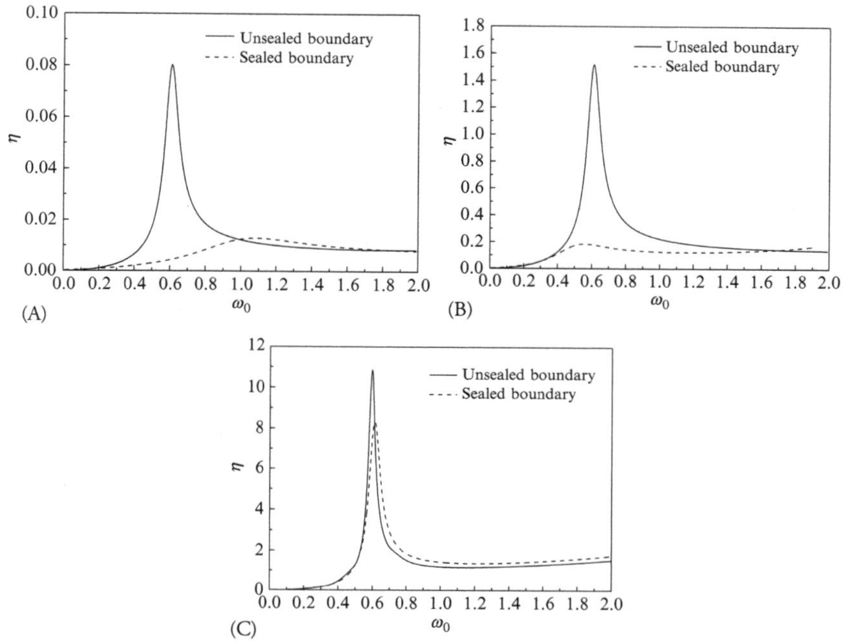

Fig. 5.1.8 Rocking displacement varied with ω_0 under different drainage conditions. (A) excited by P waves, (B) excited by SV waves, and (C) excited by Rayleigh waves

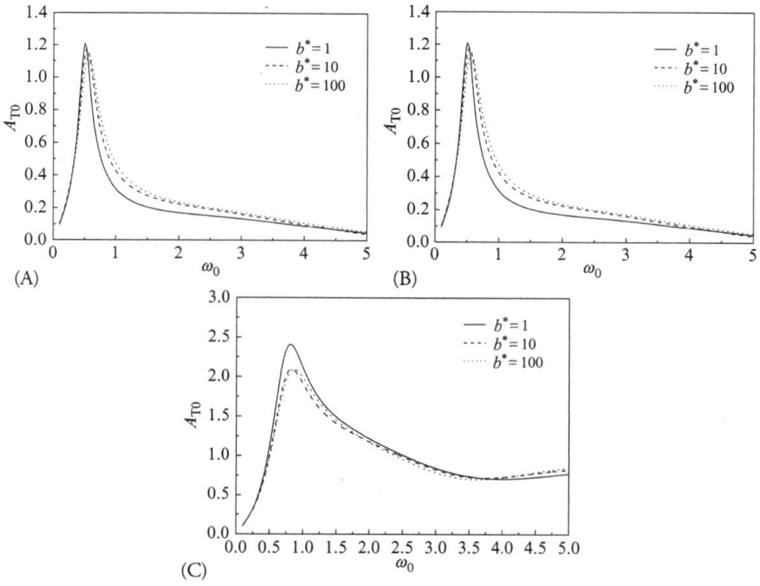

Fig. 5.1.9 Vertical displacement varied with ω_0 under different b^*. (A) excited by P waves, (B) excited by SV waves, and (C) excited by Rayleigh waves

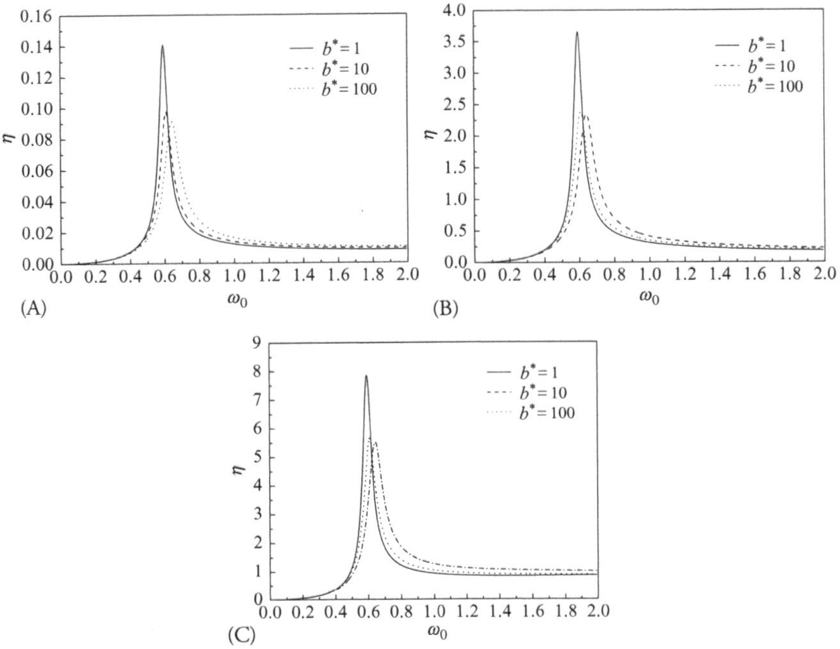

Fig. 5.1.10 Rocking displacement varied with ω_0 under different b^*. (A) excited by P waves, (B) excited by SV waves, and (C) excited by Rayleigh waves

The incident angle is another key factor that directly affects the displacement of the foundation excited by incident waves. Figs. 5.1.11 and 5.1.12 display the variation of vertical displacement with incident angle under different internal frictions. The excitation frequency is $\omega_0 = 0.5$, and the dimensionless mass of the foundation is $m^* = 10$. An angle of $\theta_{\rm in} = 0°$ corresponds to the vertical incidence, while $\theta_{\rm in} = 90°$ corresponds to the horizontal incidence. If the incident waves are in the case of vertical incidence or horizontal incidence, there is no rocking vibration to be excited (Fig. 5.1.12). The same regulation is obtained for vertical vibration under SV waves (Fig. 5.1.12B). For vertical vibration under SV waves, the maximum vertical displacement occurs at vertical incidence, while the displacement corresponding to horizontal incidence is zero. It can also be seen that the displacement of poroelastic soil decreases as the intrinsic permeability increases. In addition, oscillations of soil vertical displacement are observed when the incident angles are in the range of $15° < \theta_{\rm in} < 45°$, as shown in Figs. 5.1.11B and 5.1.12B. According to Lin et al.[34], for incident SV waves, the coefficients of reflected P waves oscillate dramatically when the incident angle approaches the critical angle. It therefore results in the fluctuations of soil vertical displacement.

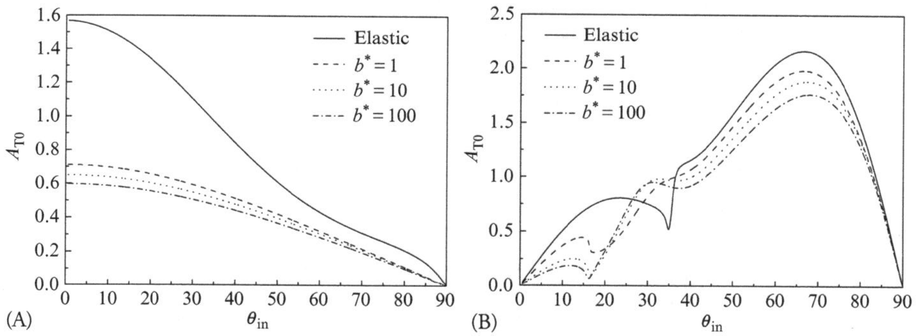

Fig. 5.1.11 Vertical displacement varied with θ_{in} under different b^*. (A) excited by P waves and (B) excited by SV waves

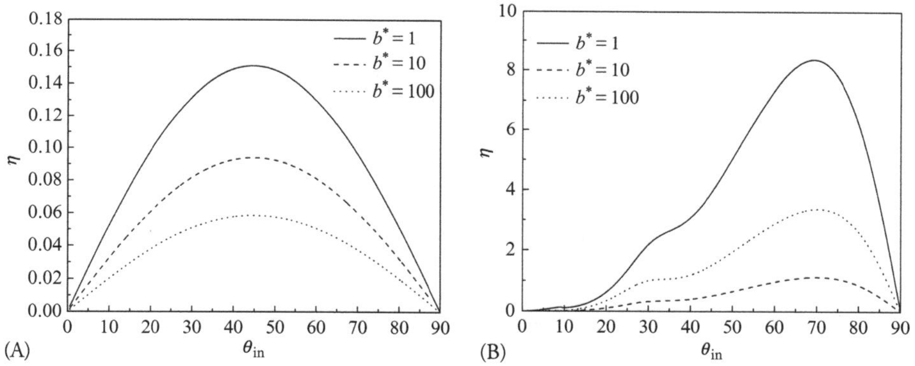

Fig. 5.1.12 Rocking displacement varied with θ_{in} under different b^*. (A) excited by P waves and (B) excited by SV waves

5.2 Dynamic Response of a Rigid Foundation in Saturated Soil to Plane Waves

5.2.1 Structural Model

The foundation-soil system considered corresponds to a rigid circular foundation embedded in a uniform poroelastic half-space. Nonvertically incident P and SV waves with time dependence $e^{i\omega t}$ act on the foundation, in which ω denotes frequency. The geometry of the foundation and the coordinate system are depicted in Fig. 5.2.1.

The influence of the embedment depth of the foundation was analyzed based on the procedure presented by Novak et al.[41–43] and some similar assumptions were introduced:

All the motions studied here were assumed to be time-harmonic and characterized by the same time dependence $e^{i\omega t}$, which will be omitted from now on for simplicity.

The soil underlying the foundation is a homogeneous poroelastic half-space. The dynamic reaction at the base of the foundation was independent of the embedment depth and the overlying soil.

The soil along the vertical side of the foundation was composed of a series of infinitesimally thin poroelastic layers. And in each layer, the gradient of the normal stress component in the vertical direction was neglected.

The soil and the foundation were assumed to be perfectly bonded, and the surface between them was smooth and fully permeable.

The governing model of the soil and the total wave fields was the same as mentioned in Sections 5.1.1 and 5.1.2.

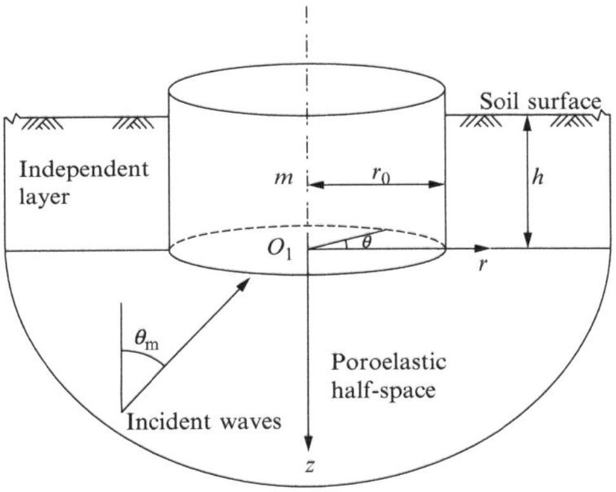

Fig. 5.2.1 Rigid foundation in poroelastic half-space and subjected to incident waves

5.2.2 The Vertical Vibration of the Foundation

The Force Acting on the Base of the Foundation

The total force caused by rigid-body scattering waves and radiation scattering waves acting on the foundation base is obtained by employing the same procedure as in Section 5.1.2, which has the following form:

$$T_1^* = \frac{4A_{\mathrm{T0}}}{L}\int_0^1 \theta_{\mathrm{R}}(x)\mathrm{d}x + \frac{4}{L}\int_0^1 \theta_{\mathrm{S}}(x)\mathrm{d}x \tag{5.2.1}$$

Parameters in Eq. (5.2.1) can be found in Section 5.1.2.

The total force acting on the foundation base caused by free waves is

$$
\begin{aligned}
T_2^* &= \int_0^{2\pi}\int_0^1 \sigma_{zz}(r^*,\theta,h^*)r^*\mathrm{d}r^*\mathrm{d}\theta \\
&= -2\pi(\lambda_c+\alpha\xi_1 M)k_{p1}^2\left(a_0\mathrm{e}^{\mathrm{i}k_{p1z}h^*}+a_1\mathrm{e}^{-\mathrm{i}k_{p1z}h^*}\right)\int_0^1 J_0(k_e r^*)r^*\mathrm{d}r^* \\
&\quad -2\pi(\lambda_c+\alpha\xi_2 M)k_{p2}^2 a_2\mathrm{e}^{-\mathrm{i}k_{p2z}h^*}\int_0^1 J_0(k_e r^*)r^*\mathrm{d}r^* \\
&\quad +4\pi\left[-a_0k_{p1z}^2\mathrm{e}^{\mathrm{i}k_{p1z}h^*}-a_1k_{p1z}^2\mathrm{e}^{-\mathrm{i}k_{p1z}h^*}-a_2k_{p2z}^2\mathrm{e}^{-\mathrm{i}k_{p2z}h^*}\right]\int_0^1 J_0(k_e r^*)r^*\mathrm{d}r^* \\
&\quad -\frac{4\mathrm{i}\pi k_e^2 k_{sz}b_1}{k_s}\int_0^1 J_0(k_e r^*)r^*\mathrm{d}r^*
\end{aligned}
\tag{5.2.2}
$$

The Force Acting on the Side of the Foundation

Following the previous assumptions, governing equations for the soil along the vertical side of the foundation can be rewritten as

$$\mu\frac{\partial^2 u_z}{\partial r^2}+\mu\frac{1}{r}\frac{\partial u_z}{\partial r}=\rho\ddot{u}_z+\rho_f\ddot{w}_z \tag{5.2.3}$$

$$0=\rho_f\ddot{u}_z+\frac{\rho_f}{n}\ddot{w}_z+b\dot{w}_z \tag{5.2.4}$$

$$\sigma_{rz}=\mu\frac{\partial u_z}{\partial r} \tag{5.2.5}$$

and the boundary condition of the interface between the soil and the side of the foundation is

$$u_z^{\mathrm{R}}(z)=U_z\ (r=r_0) \tag{5.2.6}$$

$$u_z^{\mathrm{S}}(z)=-u_z^{\mathrm{in+re}}(z)\ (r=r_0) \tag{5.2.7}$$

Introducing the dimensionless parameters and applying Hankel transforms, the solutions of Eqs. (5.2.3)–(5.2.5) can be expressed as follows:

$$u_z(r^*,z^*)=AH_0^2(r^*D_3) \tag{5.2.8}$$

$$\sigma_{rz}(r^*,z^*)=-D_3AH_1^2(r^*D_3) \tag{5.2.9}$$

where A is an undetermined coefficient, H_0^2, H_1^2 are the second kind of the zero-order and first-order Hankel functions, $D_3=\omega_0\sqrt{1+\frac{n\rho^*\omega_0}{\mathrm{i}nb^*-\rho^*\omega_0}}$, and $\mathrm{Re}(D_3)>0$.

Combining with the boundary condition of Eqs. (5.2.6), (5.2.7), the force caused by rigid-body scattering waves and radiation scattering waves acting on the side of the foundation can be obtained by the integration of σ_{rz} along the circumference of the side of the foundation:

$$T_3^* = -2\pi h^* A_{\mathrm{T0}} D_1 \frac{H_1^2(D_3)}{H_0^2(D_3)} \tag{5.2.10}$$

$$T_4^* = \left(a_0 \mathrm{e}^{\mathrm{i}k_{\mathrm{p1z}}h^*} + a_1 \mathrm{e}^{-\mathrm{i}k_{\mathrm{p1z}}h^*} + a_2 \mathrm{e}^{-\mathrm{i}k_{\mathrm{p2z}}h^*} - \frac{b_1 k_\mathrm{e}^2}{\mathrm{i}k_\mathrm{s}^2} \mathrm{e}^{-\mathrm{i}k_{\mathrm{sz}}h^*} - a_0 - a_1 - a_2 + \frac{b_1 k_\mathrm{e}^2}{\mathrm{i}k_\mathrm{s}^2} \right) \cdot 2\pi D_1 J_0(k_\mathrm{e}) \frac{H_1^2(D_1)}{H_0^2(D_1)} \tag{5.2.11}$$

The total force acting on the side of the foundation caused by free waves is

$$T_5^* = \left(a_0 \mathrm{e}^{\mathrm{i}k_{\mathrm{p1z}}h^*} + a_1 \mathrm{e}^{-\mathrm{i}k_{\mathrm{p1z}}h^*} + a_2 \mathrm{e}^{-\mathrm{i}k_{\mathrm{p2z}}h^*} - \frac{b_1 \left(k_\mathrm{e}^2 - k_{\mathrm{sz}}^2\right)}{2\mathrm{i}k_\mathrm{s}^2} \left(\mathrm{e}^{-\mathrm{i}k_{\mathrm{sz}}h^*} - 1 \right) - a_0 - a_1 - a_2 \right) \cdot 4\pi k_\mathrm{e} J_1(k_\mathrm{e}) \tag{5.2.12}$$

Dynamic Equilibrium of the Foundation

Introducing the dynamic equilibrium equation of the foundation and combining with Eqs. (5.2.1), (5.2.2), (5.2.10)–(5.2.12), the following equation can be obtained:

$$-\omega_0^2 m^* A_{\mathrm{T0}} = \frac{4A_{\mathrm{T0}}}{L} \int_0^1 \theta_\mathrm{R}(x)\mathrm{d}x + \frac{4}{L} \int_0^1 \theta_\mathrm{S}(x)\mathrm{d}x + T_2^* - A_{\mathrm{T0}} D_3 \frac{H_1^2(D_3)}{H_0^2(D_3)} + T_4^* + T_5^* \tag{5.2.13}$$

where m^* is the dimensionless mass of the foundation, $m^*=m/\rho r_0^5$.

We introduce the definition

$$K_{\mathrm{TT}} = \frac{4}{L} \int_0^1 \theta_\mathrm{R}(x)\mathrm{d}x - D_3 \frac{H_1^2(D_3)}{H_0^2(D_3)} \tag{5.2.14}$$

According to the vibration analysis method presented by Richart[40], the vertical displacement amplitude of the foundation excited by incident P1 wave can be written as

$$A_{\mathrm{T0}} = \frac{\left| \frac{4}{L} \int_0^1 \theta_\mathrm{S}(t)\mathrm{d}t + T_4^* + T_5^* \right|}{\sqrt{\left(k_\mathrm{t} - m^* \omega_0^2\right)^2 + \left(c_\mathrm{t}\omega_0\right)^2}} \tag{5.2.15}$$

where k_t =-Re(K_{TT}), c_t=-Im(K_{TT})/ω_0.

5.2.3 The Rocking Vibration of the Foundation

The Force Acting on the Base of the Foundation

The total force caused by rigid-body scattering waves and radiation scattering waves acting on the base of the foundation is obtained by employing the same procedure as in Section 5.1.3, which has the following form:

$$M_1^* = \frac{8\eta}{L}\int_0^1 x\theta_{\mathrm{R}}(x)\mathrm{d}x + \frac{8}{L}\int_0^1 x\theta_{\mathrm{S}}(x)\mathrm{d}x \tag{5.2.16}$$

Parameters in Eq. (5.2.16) can be found in Section 5.1.3.

The total moment acting on the base of the foundation caused by free waves is

$$\begin{aligned}
M_2^* &= \int_0^{2\pi}\int_0^1 \sigma_{zz}(r^*, \theta, h^*) r^{*2}\cos\theta \mathrm{d}r^* \mathrm{d}\theta \\
&= -\pi(\lambda_c + \alpha\xi_1 M)k_{p1}^2\left(a_0\mathrm{e}^{\mathrm{i}k_{p1z}h^*} + a_1\mathrm{e}^{-\mathrm{i}k_{p1z}h^*}\right)\int_0^1 J_1(k_e r^*) r^{*2}\mathrm{d}r^* \\
&\quad -\pi(\lambda_c + \alpha\xi_2 M)k_{p2}^2 a_2\mathrm{e}^{-\mathrm{i}k_{p2z}h^*}\int_0^1 J_1(k_e r^*) r^{*2}\mathrm{d}r^* \\
&\quad + 2\pi\left[-a_0 k_{p1z}^2\mathrm{e}^{\mathrm{i}k_{p1z}h^*} - a_1 k_{p1z}^2\mathrm{e}^{-\mathrm{i}k_{p1z}h^*} - a_2 k_{p2z}^2\mathrm{e}^{-\mathrm{i}k_{p2z}h^*}\right]\int_0^1 J_1(k_e r^*) r^{*2}\mathrm{d}r^* \\
&\quad - \frac{2\mathrm{i}\pi k_e^2 k_{sz} b_1}{k_s}\int_0^1 J_1(k_e r^*) r^{*2}\mathrm{d}r^*
\end{aligned} \tag{5.2.17}$$

The Force Acting on the Side of the Foundation

Following the previous assumptions, the governing equations of the soil along the vertical side of the foundation for a rocking vibration are

$$\mu\left(\frac{\partial^2}{\partial r^2} + \frac{1}{r}\frac{\partial}{\partial r} + \frac{1}{r^2}\frac{\partial^2}{\partial\theta^2}\right)u_z = \rho\ddot{u}_z + \rho_f\ddot{w}_z \tag{5.2.18}$$

$$0 = \rho_f\ddot{u}_z + \frac{\rho_f}{n}\ddot{w}_z + b\dot{w}_z \tag{5.2.19}$$

$$\sigma_{rz} = \mu\frac{\partial u_z}{\partial r} \tag{5.2.20}$$

Introducing the dimensionless parameters and applying Hankel transforms, the solutions of Eqs. (5.2.18)–(5.2.20) can be expressed as follows:

$$u(r^*, \theta, z^*) = \sum_{n=0}^{\infty} B_n H_1^2(D_3)\cos n\theta \tag{5.2.21}$$

$$\sigma_{rz}(r^*, \theta, z^*) = \sum_{n=0}^{\infty} B_n\left(D_1 H_0^2(D_3) - H_1^2(D_3)\right)\cos n\theta \tag{5.2.22}$$

and the corresponding boundary conditions along the contact surface are

$$\sum_{n=0}^{\infty} u_{zn}^{\mathrm{R}}(z^*)\cos n\theta = \eta r\cos\theta \quad (r^* = 1) \tag{5.2.23}$$

$$\sum_{n=0}^{\infty} u_{zn}^{\mathrm{S}}(h^*)\cos n\theta = -u_z^{\mathrm{in+re}}(z^*) \quad (r^* = 1) \tag{5.2.24}$$

Then, the moments caused by rigid-body scattering waves and radiation scattering waves acting on the side of the foundation can be obtained as follows:

$$M_3^* = \eta\pi h^* \left[\frac{D_3 H_0^2(D_3)}{H_1^2(D_1)} - 1\right] \tag{5.2.25}$$

$$\begin{aligned} M_4^* = &\left(a_0 \mathrm{e}^{\mathrm{i}k_{\mathrm{p1z}}h^*} + a_1 \mathrm{e}^{-\mathrm{i}k_{\mathrm{p1z}}h^*} + a_2 \mathrm{e}^{-\mathrm{i}k_{\mathrm{p2z}}h^*} - \frac{b_1 k_{\mathrm{e}}^2}{\mathrm{i}k_{\mathrm{s}}k_{\mathrm{sz}}} \mathrm{e}^{-\mathrm{i}k_{\mathrm{sz}}h^*} - a_0 - a_1 - a_2 + \frac{b_1 k_{\mathrm{e}}^2}{\mathrm{i}k_{\mathrm{s}}k_{\mathrm{sz}}}\right) \\ &\cdot 2\mathrm{i}\pi J_1(k_{\mathrm{e}}) \left[D_3 \frac{H_0^2(D_3)}{H_1^2(D_3)} - 1\right] \end{aligned} \tag{5.2.26}$$

The moment acting on the side of the foundation caused by free waves is

$$\begin{aligned} M_5^* = &-\left(a_0 \mathrm{e}^{\mathrm{i}k_{\mathrm{p1z}}h^*} + a_1 \mathrm{e}^{-\mathrm{i}k_{\mathrm{p1z}}h^*} + a_2 \mathrm{e}^{-\mathrm{i}k_{\mathrm{p2z}}h^*} - \frac{b_1 \left(k_{\mathrm{e}}^2 - k_{\mathrm{sz}}^2\right)}{2\mathrm{i}k_{\mathrm{s}}k_{\mathrm{sz}}} \left(\mathrm{e}^{-\mathrm{i}k_{\mathrm{sz}}h^*} - 1\right) - a_0 - a_1 - a_2\right) \\ &\cdot 4\mathrm{i}\pi [J_1(k_{\mathrm{e}}) - k_{\mathrm{e}} J_2(k_{\mathrm{e}})] \end{aligned} \tag{5.2.27}$$

Dynamic Equilibrium of the Foundation

Combining with Eqs. (5.2.16), (5.2.17), (5.2.25)–(5.2.27), the dynamic equilibrium equation of the foundation implies

$$-\omega_0^2 m^* \eta = \frac{8\eta}{L}\int_0^1 \theta_{\mathrm{R}}(x)\mathrm{d}x + \frac{4}{L}\int_0^1 \theta_{\mathrm{S}}(x)\mathrm{d}x + M_2^* + \eta\pi h^* \left[\frac{D_3 H_0^2(D_3)}{H_1^2(D_3)} - 1\right] + M_4^* + M_5^* \tag{5.2.28}$$

Introducing the definition

$$K_{\mathrm{TT}} = \frac{8\eta}{L}\int_0^1 \theta_{\mathrm{R}}(x)\mathrm{d}x + \eta\pi h^* \left[\frac{D_3 H_0^2(D_3)}{H_1^2(D_3)} - 1\right] \tag{5.2.29}$$

Finally, the rocking displacement amplitude of the foundation to P waves is

$$\eta = \frac{\left|\dfrac{8}{\eta}\displaystyle\int_0^1 \theta_{\mathrm{S}}(t)\mathrm{d}t + M_4^* + M_5^*\right|}{\sqrt{(k_{\mathrm{t}} - m^*\omega_0^2)^2 + (c_{\mathrm{t}}\omega_0)^2}} \tag{5.2.30}$$

where $k_{\mathrm{t}} = -\mathrm{Re}(K_{\mathrm{TT}})$, $c_{\mathrm{t}} = -\mathrm{Im}(K_{\mathrm{TT}})/\omega_0$.

Employing the same procedure, the vibrations of the embedded foundation excited by incident SV waves can be obtained, and this is not repeated here for the purpose of brevity.

5.2.4 Numerical Results and Discussion

Effects of excitation frequency and depth of the embedment on foundation vibration are given in Figs.

5.2.2 and 5.2.3, in which the incident angle is $\theta_{in} = 30°$ and $b^* = 10$. Because the dynamic reaction at the base of the foundation is independent of the overlying soil in the present model suggested by Novak, the difference between numerical results and practical situations becomes large when the depth of the embedment increases. In this study, four different values of the embedded depth are considered: $h^* = 0, 0.25, 0.5, 0.75$. As can be seen, for both vertical vibration and rocking vibration, there is resonance. And a comparison between Figs. 5.2.2 and 5.2.3 indicates that the resonance of the rocking vibration is more significant.

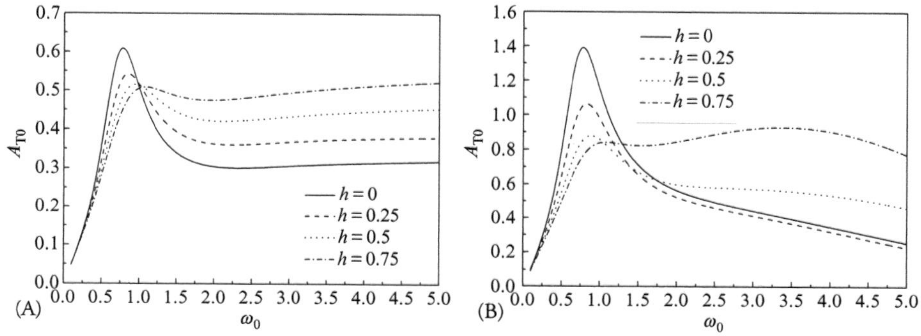

Fig. 5.2.2 The influence of h^* on the vertical vibration of the foundation. (A) by P waves and (B) by SV waves

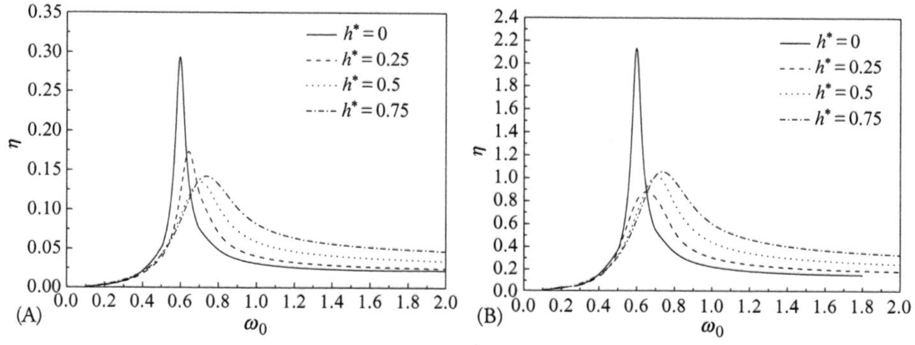

Fig. 5.2.3 The influence of h^* on the rocking vibration of the foundation. (A) By P waves and (B) by SV waves

It can also be noticed that the resonance amplitudes of the displacement curves decrease with the increasing of h^*. At the same time, the increasing of h^* weakens the resonance phenomenon of the foundation. That is, the displacement curves decay more slowly for a larger h^* with the increasing of ω_0 from the resonant frequency, such as $\omega_0 > 1$ in Fig. 5.2.2 and $\omega_0 > 0.75$ in Fig. 5.2.3. This phenomenon is more obvious for vertical vibration. In particular, when $h^* = 0.75$ as in Fig. 5.2.2, the attenuation of the displacement curve does not occur. And for the rocking vibration in Fig. 5.2.3, the increase in h^* leads simultaneously to the increase in resonance frequency.

Figs. 5.2.4 and 5.2.5 present the influence of the mass of the foundation on the embedded foundation vibration, in which h^* = 0.5, b^* = 10 and the incident angle is θ_{in} = 30°. It is noted that the influence of m^* on the embedded foundation vibration is the same as that for the surface foundation. As there is an increase in ω_0, the curve becomes sensitive to the mass of the foundation. As can be seen, when ω_0 > 0.75, the displacements decrease and the resonant frequency shifts to a lower value as there is an increase in m^*. But for the rocking vibration shown in Fig. 5.2.5, differences between curves of m^* = 15 and m^* = 20 became less when ω_0 > 1.0.

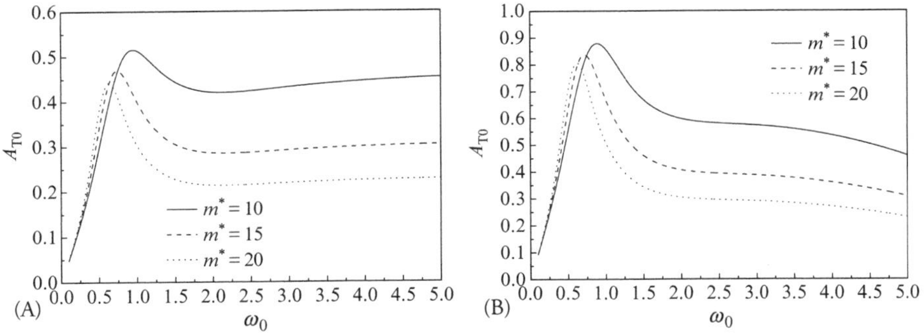

Fig. 5.2.4 The influence of m^* on the vertical vibration of the foundation. (A) by P waves and (B) by SV waves

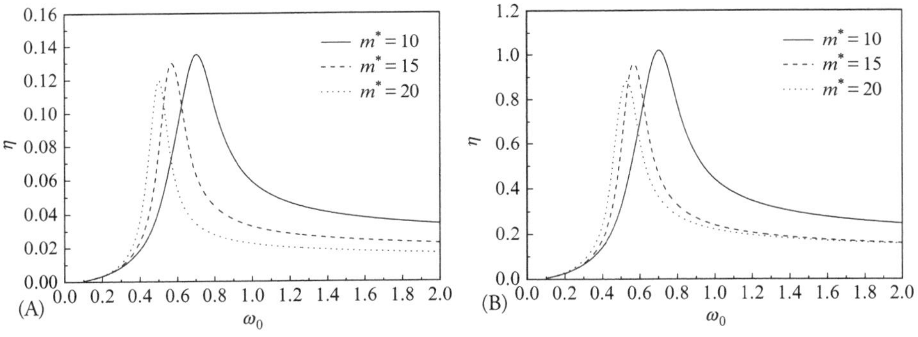

Fig. 5.2.5 The influence of m^* on the rocking vibration of the foundation. (A) by P waves and (B) by SV waves

The vibration and rocking displacements of the embedded foundation varied with ω_0 for different b^* as shown in Fig. 5.2.6. Here, h^* = 0.5, m^* = 10 and the incident angle is θ_{in} = 30°. It can be noted that the differences in the curves under different b^* are quite small, and can be neglected when ω_0 is smaller than the resonant frequency. When ω_0 is larger than the resonant frequency, the displacements increase with the increase in b^*. For the vertical vibration, the displacements decrease from the resonant amplitude, and then increase slightly with the increase in ω_0. Inflection points of curves are observed at about ω_0 = 2 in Fig. 5.2.6. The same phenomenon is not observed in Fig. 5.2.7, but the resonant amplitudes of the rocking vibration decrease a little when b^* increases, which is not shown in Fig. 5.2.6.

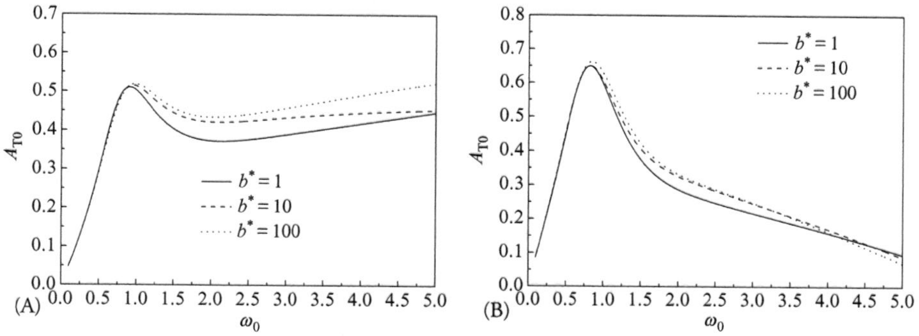

Fig. 5.2.6 The influence of b^* on the vertical vibration of the foundation. (A) by P waves and (B) by SV waves

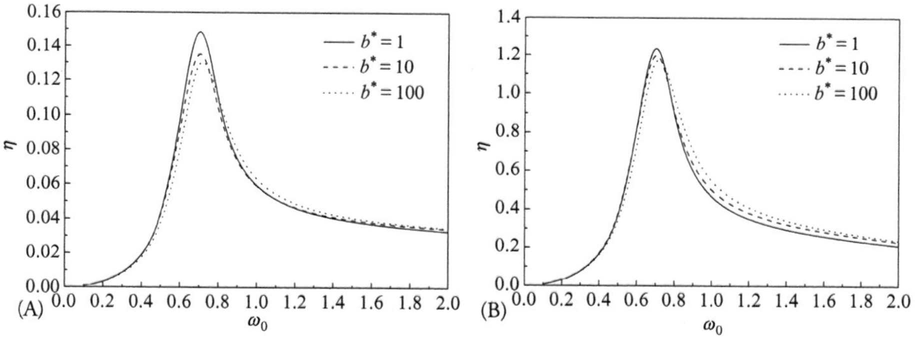

Fig. 5.2.7 The influence of b^* on the rocking vibration of the foundation. (A) by P waves and (B) by SV waves

As shown in Fig. 5.2.1, we introduced the definition that θ_{in} = 0° corresponds to vertical incidence, while θ_{in} = 90° corresponds to horizontal incidence. Here, three different values of the incident angle were considered: θ_{in} = 30°, 45°, and 60°, respectively. The influence of θ_{in} on the vibration was obvious. For both vertical and rocking vibration, different θ_{in} values led to a different resonant amplitude. The displacements of the vertical vibration under P waves decreased as θ_{in} increased from 30° to 60°, accompanied by the decrease in the resonant frequency in Fig. 5.2.8A. And for the vertical vibration under SV waves in Fig. 5.2.8B and for the rocking vibration in Fig. 5.2.9, the resonant frequency stays constant for different θ_{in} values. However, for the vertical vibration under SV Fig. 5.2.8B, the curve fluctuates with ω_0 and a minimum point can be found at around ω_0 = 3.4.

Figs. 5.2.10 and 5.2.11 show the variation of the displacements with θ_{in}. Considering the importance of the resonance phenomenon in practical conditions, the dynamic response was assumed to arise at ω_0 = 0.75, which approximately equals the resonance frequency, according to Figs. 5.2.8 and 5.2.9. The same regulations as those for the surface foundation were found and the influence of the critical angle of SV waves was also observed.

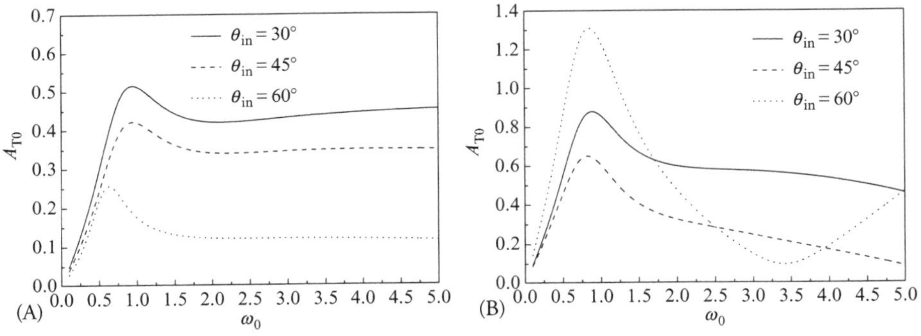

Fig. 5.2.8 The influence of θ_{in} on the vertical vibration of the foundation. (A) by P waves and (B) by SV waves

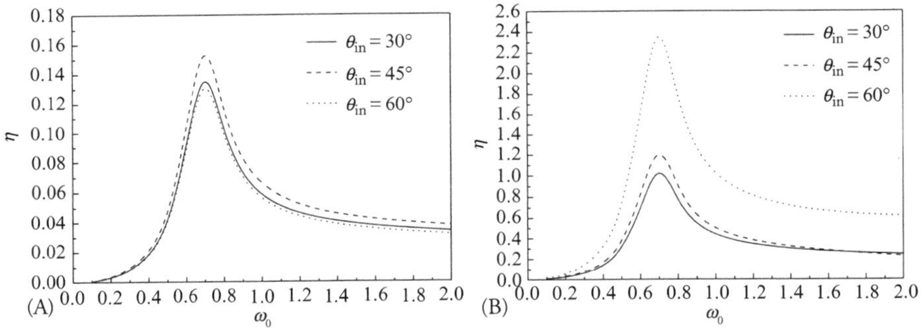

Fig. 5.2.9 The influence of θ_{in} on the rocking vibration of the foundation. (A) by P waves and (B) by SV waves

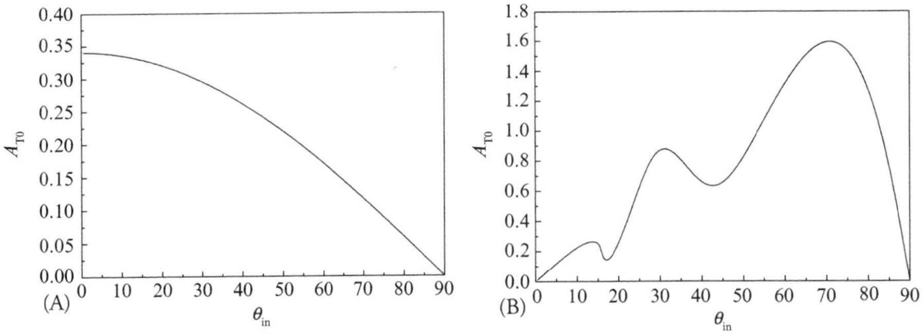

Fig. 5.2.10 The vertical vibration varies with θ_{in}. (A) by P waves and (B) by SV waves

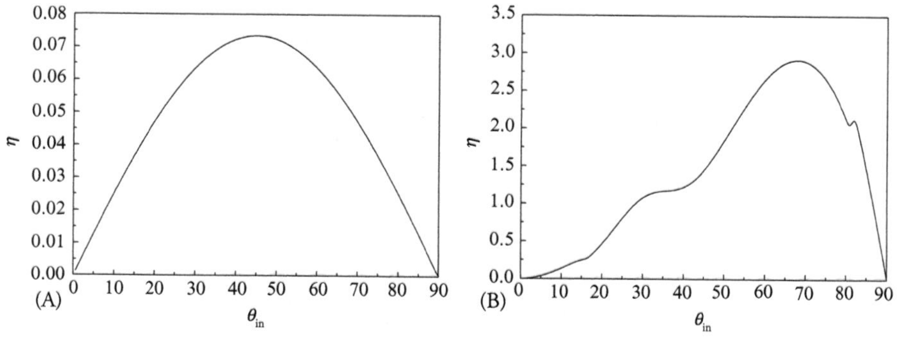

Fig. 5.2.11 The rocking vibration varies with θ_{in}. (A) by P waves and (B) by SV waves

5.3 Conclusions

The dynamic response of a circular rigid foundation on a poroelastic medium excited by incident plane waves was investigated based on Biot's dynamic poroelastic theory. The following conclusions can be drawn:

(1) The resonant phenomenon arises regardless of vertical vibration or rocking vibration. Both the resonant displacement and resonant frequency decrease as the foundation mass increases.

(2) Pore water imposes a significant effect on the vibrations. The peak value of the vertical displacement of the foundation on a poroelastic half-space is smaller than that for an elastic medium. And the displacement of an unsealed boundary is larger than that of a sealed boundary.

(3) The maximum vertical displacements occur at vertical incidence for P waves, while the displacements under horizontal incidence are zero. The vertical displacements for SV waves and the rocking displacements for both P and SV waves generally increase from zero at the beginning and decay to zero as the incident angle increases. The displacement curves show fluctuations when the incident angle is equal to the critical angle or total reflection occurs.

(4) For both vertical vibration and rocking vibration, the resonance amplitude of the embedded foundation is less than that of the surface foundation. As the embedded depth increases, the vertical displacements increase significantly when ω_0 is larger than the resonance frequency.

Chapter 6
Isolation of Elastic Waves

The special focus on environmental issues, such as the adverse effects of the vibrations generated by traffic loads, machine foundations, and pile driving or construction blasting, has led to an increasing interest in methods for isolating vibrations. In order to impede the transmission of vibrations in soil and reduce the unfavorable effects of soil oscillations upon inhabitants and structures, different types of wave barriers can be used, varying from stiff concrete walls or piles to flexible gas cushions. In this chapter, a row of piles will be addressed as a countermeasure for isolating elastic waves in poroelastic soil. A parametric study is conducted to investigate the influence of certain characteristics of poroelastic soil and piles, such as the permeability of poroelastic soil, the pile spacing and the number of piles, on the isolation performance of a pile barrier.

6.1 Isolation of Elastic Plane Waves in Poroelastic Soil

6.1.1 Governing Equations for Poroelastic Soil

The constitutive equations and motion-governing equations for homogeneous poroelastic soil can be expressed as in Eqs. (1.1.21), (1.1.22) and (1.1.24), (1.1.25).

The Helmholtz decomposition theorem allows one to resolve the governing Eqs. (1.1.24), (1.1.25) in terms of the potentials φ, ψ, χ, and Θ, as follows:

$$u=\text{grad}\ \varphi+\text{curl}\ \psi \tag{6.1.1}$$

$$w=\text{grad}\ \chi+\text{curl}\ \Theta \tag{6.1.2}$$

In the steady-state case, the soil skeleton and fluid displacements are the harmonic functions of the angular frequency ω. Since the factor $\mathrm{e}^{-\mathrm{i}\omega t}$ is shared by all the variables (excitations and responses), it will be omitted from now on for simplicity. Thus, omitting the time dependency, and substituting Eqs. (6.1.1), (6.1.2) into Biot's field equations of motion, Eqs. (1.1.24), (1.1.25), two sets of coupled equations are obtained as

$$\begin{bmatrix} \lambda_c+\alpha^2 M & \alpha M \\ \alpha M & M \end{bmatrix}\begin{bmatrix} \nabla^2\varphi \\ \nabla^2\chi \end{bmatrix}=\begin{bmatrix} -\rho\omega^2 & -\rho_f\omega^2 \\ -\rho_f\omega^2 & -h\omega^2-\mathrm{i}\omega b \end{bmatrix}\begin{bmatrix} \varphi \\ \chi \end{bmatrix} \tag{6.1.3}$$

$$\begin{bmatrix} \mu & 0 \\ 0 & \mu \end{bmatrix} \begin{bmatrix} \nabla^2 \psi \\ \nabla^2 \Theta \end{bmatrix} = \begin{bmatrix} -\rho\omega^2 & -\rho_f\omega^2 \\ -\rho_f\omega^2 & -h\omega^2 - i\omega b \end{bmatrix} \begin{bmatrix} \psi \\ \Theta \end{bmatrix} \tag{6.1.4}$$

where $i = \sqrt{-1}$; $\lambda_c = \lambda + \alpha^2 M$; ∇^2 is the Laplacian operator.

By eliminating χ from Eq. (6.1.3), the resulting equation governing φ is then reducible to

$$A\nabla^4\varphi + B\nabla^2\varphi + C\varphi = 0 \tag{6.1.5}$$

where

$$A = (\lambda + 2\mu)M, \; B = (\lambda_c + 2\mu)(h\omega^2 + i\omega b) + \rho\omega^2 M - 2\rho_f\,\omega^2\alpha M$$

$$C = \rho\omega^2\,(h\omega^2 + i\omega b) - \rho_f^2\omega^4 \tag{6.1.6}$$

The solution of Eq. (6.1.5) may be written in the form

$$\varphi = \varphi_f + \varphi_s \tag{6.1.7}$$

where

$$(\nabla^2 + p_{1,2}^2)\varphi_{f,s} = 0 \tag{6.1.8}$$

$$p_{1,2}^2 = \frac{B \mp \sqrt{B^2 - 4AC}}{2A} \tag{6.1.9}$$

and where p_1 and p_2 designate the complex wave numbers of the fast and slow compressional waves, respectively.

With the aid of Eqs. (6.1.7), (6.1.8), the remaining potential χ in Eq. (6.1.3) is then found to be

$$\chi = \xi_1\,\varphi_f + \xi_2\varphi_s \tag{6.1.10}$$

where the amplitude ratios ξ_1 and ξ_2 are given by

$$\xi_{1,2} = \frac{(\lambda + \alpha^2 M + 2\mu)p_{1,2}^2 - \rho\omega^2}{\rho_f\omega^2 - \alpha M p_{1,2}^2} \tag{6.1.11}$$

Similarly, by eliminating Θ from Eq. (6.1.4), the resulting equation governing ψ is reducible to

$$(\nabla^2 + p_3^2)\psi = 0 \tag{6.1.12}$$

where p_3 denotes the complex wave number of the transverse waves, and is given by

$$p_3^2 = \frac{C}{D} \tag{6.1.13}$$

where

$$D=\mu(h\omega^2+\mathrm{i}b\omega) \tag{6.1.14}$$

Considering Eqs. (6.1.7), (6.1.8), the remaining potential Θ in Eq. (6.1.3) can be written as

$$\Theta=\xi_3\psi \tag{6.1.15}$$

where

$$\xi_3 = -\frac{\rho_{\mathrm{f}}\omega^2}{h\omega^2 + \mathrm{i}b\omega} \tag{6.1.16}$$

6.1.2 Isolation of Plane Waves by Pile Rows

Incidence of Transverse Plane Waves

Fig. 6.1.1 shows a transverse plane wave of amplitude ψ_0 incident at an angle θ_β on a row of elastic piles in a solid matrix. Such an incident wave is represented in the reference system (x_1, y_1) attached to the first pile by means of

$$\psi^{(\mathrm{inc})}(x_1, y_1)=\psi_0 \mathrm{e}^{\mathrm{i}p3\ (x_1 \cos\theta_\beta+y_1 \sin\theta_\beta)} \tag{6.1.17}$$

In the presence of a row of elastic piles, the incident waves will be scattered and diffracted around the piles. The solutions are therefore given by

$$\varphi_1 = \sum_{k=1}^{N} \varphi_{\mathrm{f}}^k(r_k, \theta_k) + \sum_{k=1}^{N} \varphi_{\mathrm{s}}^k(r_k, \theta_k) \tag{6.1.18}$$

$$\psi_1 = \psi^{(\mathrm{inc})}(x_1, y_1) + \sum_{k=1}^{N} \psi^k(r_k, \theta_k) \tag{6.1.19}$$

where $\varphi_{\mathrm{f}}^k(r_k, \theta_k)$, $\varphi_{\mathrm{s}}^k(r_k, \theta_k)$, and $\psi_k(r_k, \theta_k)$ represent the waves scattered in a solid matrix by the kth pile in terms of the kth system of cylindrical coordinates (r_k, θ_k); N is the number of piles.

Using the method of separation of variables, the scattered field in a solid matrix by the kth pile in the reference system (r_k, θ_k) can be written as

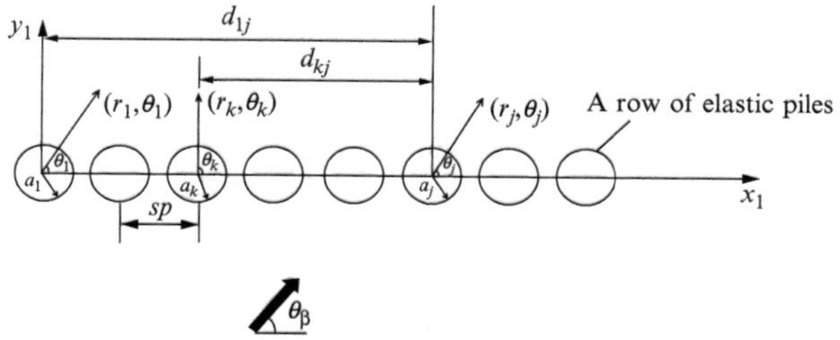

Fig. 6.1.1 Incident plane waves and reference systems for each pile

$$
\begin{aligned}
\varphi_{\mathrm{f}}^{k}(r_k,\theta_k) &= \sum_{n=0}^{\infty} H_n^{(1)}(p_1 r_k)\left(A_{1n}^{k}\cos n\theta_k + B_{1n}^{k}\sin n\theta_k\right)\\
\varphi_{\mathrm{s}}^{k}(r_k,\theta_k) &= \sum_{n=0}^{\infty} H_n^{(1)}(p_2 r_k)\left(C_{1n}^{k}\cos n\theta_k + D_{1n}^{k}\sin n\theta_k\right)\\
\psi^{k}(r_k,\theta_k) &= \sum_{n=0}^{\infty} H_n^{(1)}(p_3 r_k)\left(E_{1n}^{k}\sin n\theta_k + F_{1n}^{k}\cos n\theta_k\right)
\end{aligned}
\tag{6.1.20}
$$

where $H_n^{(1)}(\cdot)$ is the Hankel function of the first kind, and order n, representing an outgoing wave that satisfies the Sommerfeld radiation condition at infinity; A_{1n}^k, B_{1n}^k, C_{1n}^k, D_{1n}^k, E_{1n}^k, and F_{1n}^k are complex coefficients to be determined from boundary conditions.

In order to formulate the boundary conditions, Eqs. (6.1.18), (6.1.19) should be referred to each coordinate system (r_j, θ_j) where $1 \le j \le N$; Eq. (6.1.17) can be rewritten as

$$
\psi^{(\mathrm{inc})}(r_1,\theta_1)=\psi_0\,\mathrm{e}^{\mathrm{i}p_3 d_{1j}\cos\theta_\beta}\cdot\mathrm{e}^{\mathrm{i}p_3 r_j\cos(\theta_j-\theta_\beta)} \tag{6.1.21}
$$

where $x_j = x_1 - d_{1j}$, $y_j = y_1$; d_{1j} is the distance from the origin of system (r_1, θ_1) to the origin of system (r_j, θ_j).

Eq. (6.1.21) can be expanded in the Fourier-Bessel series as

$$
\psi^{(\mathrm{inc})}(r_1,\theta_1)=\psi_0\mathrm{e}^{\mathrm{i}p_3 d_{1j}\cos\theta_\beta}\cdot\sum_{m=0}^{\infty}\varepsilon_m \mathrm{i}^m J_m(p_3 r_j)\cos\left[m\left(\theta_j-\theta_\beta\right)\right] \tag{6.1.22}
$$

where $\varepsilon_0=1$, $\varepsilon_m=2$ when $m \ge 1$; $J_m(\cdot)$ is the Bessel function of the first kind and order m.

The wave potentials with respect to the cylindrical coordinate systems (r_j, θ_j) and (r_k, θ_k), in which $j \ne k$, are related by the following transformation:

$$
H_n^{(1)}(p_l r_k)\left\{\begin{array}{c}\cos n\theta_k\\ \sin n\theta_k\end{array}\right\}=\frac{1}{2}\sum_{m=0}^{\infty}(-1)^m\varepsilon_m J_m(p_l r_j)\left\{\begin{array}{c}K_m^n\left(p_l d_{kj}\right)\cos m\theta_j\\ L_m^n\left(p_l d_{kj}\right)\sin m\theta_j\end{array}\right\}\quad (r_j \le d_{kj}) \tag{6.1.23}
$$

for $j>k$, and

$$H_n^{(1)}(p_l r_k)\begin{Bmatrix}\cos n\theta_k\\ \sin n\theta_k\end{Bmatrix}=\frac{1}{2}(-1)^n\sum_{m=0}^{\infty}\varepsilon_m J_m(p_l r_j)\begin{Bmatrix}K_m^n(p_l d_{kj})\cos m\theta_j\\ L_m^n(p_l d_{kj})\sin m\theta_j\end{Bmatrix}\quad (r_j\le d_{kj})\tag{6.1.24}$$

for $j<k$. In Eqs. (6.1.23), (6.1.24),

$$K_m^n(p_l d_{kj})=H_{m+n}^{(1)}(p_l d_{kj})+(-1)^n \mathrm{H}_{m-n}^{(1)}(p_l d_{kj})\tag{6.1.25}$$

$$L_m^n(p_l d_{kj})=H_{m+n}^{(1)}(p_l d_{kj})+(-1)^n \mathrm{H}_{m-n}^{(1)}(p_l d_{kj})\tag{6.1.26}$$

where d_{kj} is the distance between the kth pile and the jth pile; $l=1, 2, 3$.

With the aid of Graff's addition theorem, the total wave field in a solid matrix with respect to the cylindrical coordinate system (r_j, θ_j) can be transformed to be expressed in terms of Eqs. (6.1.18), (6.1.19):

$$\begin{aligned}
\varphi_1=&(1-\delta_{j1})\sum_{k=1}^{j-1}\frac{1}{2}\sum_{n=0}^{\infty}\sum_{m=0}^{\infty}(-1)^m\varepsilon_m J_m(p_1 r_j)\\
&\cdot\left[A_{1n}^k K_m^n(p_1 d_{kj})\cos m\theta_j+B_{1n}^k L_m^n(p_1 d_{kj})\sin m\theta_j\right]\\
&+(1-\delta_{jN})\sum_{k=j+1}^{N}\frac{1}{2}\sum_{n=0}^{\infty}\sum_{m=0}^{\infty}(-1)^n\varepsilon_m J_m(p_1 r_j)\\
&\cdot\left[A_{1n}^k K_m^n(p_1 d_{kj})\cos m\theta_j+B_{1n}^k L_m^n(p_1 d_{kj})\sin m\theta_j\right]\\
&+(1-\delta_{j1})\sum_{k=1}^{j-1}\frac{1}{2}\sum_{n=0}^{\infty}\sum_{m=0}^{\infty}(-1)^m\varepsilon_m J_m(p_2 r_j)\\
&\cdot\left[C_{1n}^k K_m^n(p_2 d_{kj})\cos m\theta_j+D_{1n}^k L_m^n(p_2 d_{kj})\sin m\theta_j\right]\\
&+(1-\delta_{jN})\sum_{k=j+1}^{N}\frac{1}{2}\sum_{n=0}^{\infty}\sum_{m=0}^{\infty}(-1)^n\varepsilon_m J_m(p_2 r_j)\\
&\cdot\left[C_{1n}^k K_m^n(p_2 d_{kj})\cos m\theta_j+D_{1n}^k L_m^n(p_2 d_{kj})\sin m\theta_j\right]\\
&+\sum_{n=0}^{\infty}H_n^{(1)}(p_1 r_j)\left(A_{1n}^j\cos n\theta_j+B_{1n}^j\sin n\theta_j\right)\\
&+\sum_{n=0}^{\infty}H_n^{(1)}(p_2 r_j)\left(C_{1n}^j\cos n\theta_j+D_{1n}^j\sin n\theta_j\right)
\end{aligned}\tag{6.1.27}$$

$$\begin{aligned}\psi_1 &= \psi_0 \mathrm{e}^{\mathrm{i}p_3 d_{1j}\cos\theta_\beta}\sum_{m=0}^{\infty}\varepsilon_m \mathrm{i}^m J_m(p_3 r_j)\cos\left[m(\theta_j-\theta_\beta)\right] \\ &+(1-\delta_{j1})\cdot\sum_{k=1}^{j-1}\frac{1}{2}\sum_{n=0}^{\infty}\sum_{m=0}^{\infty}(-1)^m\varepsilon_m J_m(p_3 r_j) \\ &\cdot\left[E_{1n}^k L_m^n(p_3 d_{kj})\sin m\theta_j + F_{1n}^k K_m^n(p_3 d_{kj})\cos m\theta_j\right] \\ &+(1-\delta_{jN})\sum_{k=j+1}^{N}\frac{1}{2}\sum_{n=0}^{\infty}\sum_{m=0}^{\infty}(-1)^n\varepsilon_m J_m(p_3 r_j) \\ &\cdot\left[E_{1n}^k L_m^n(p_3 d_{kj})\sin m\theta_j + F_{1n}^k K_m^n(p_3 d_{kj})\cos m\theta_j\right] \\ &+\sum_{n=0}^{\infty}H_n^{(1)}(p_3 r_j)\left(E_{1n}^j\sin n\theta_j + F_{1n}^j\cos n\theta_j\right)\end{aligned} \tag{6.1.28}$$

where δ_{jk} is the Kronecker delta: $\delta_{jk}=1$ if $j=k$; $\delta_{jk}=0$ if $j\neq k$.

In the cylindrical coordinates (r_j, θ_j), the radial and tangential displacement components of the skeletal frame and fluid with regard to the solid and the stress components, by using the scalar potentials φ_1, ψ_1, χ_1, and Θ_1, can be derived as

$$\begin{cases} u_r = \dfrac{\partial\varphi_1}{\partial r}+\dfrac{1}{r}\dfrac{\partial\psi_1}{\partial\theta}, \quad u_\theta = \dfrac{1}{r}\dfrac{\partial\varphi_1}{\partial\theta}-\dfrac{\partial\psi_1}{\partial r}, \quad w_r = \dfrac{\partial\chi_1}{\partial r}+\dfrac{1}{r}\dfrac{\partial\Theta_1}{\partial\theta} \\ \sigma_{rr} = \lambda_c\nabla^2\varphi_1+\alpha M\nabla^2\chi_1+2\mu\left[\dfrac{\partial^2\varphi_1}{\partial r^2}+\dfrac{\partial}{\partial r}\left(\dfrac{1}{r}\dfrac{\partial\psi_1}{\partial\theta}\right)\right] \\ \tau_{r\theta} = \mu\left[2\left(\dfrac{1}{r}\dfrac{\partial^2\varphi_1}{\partial r\partial\theta}-\dfrac{1}{r^2}\dfrac{\partial\varphi_1}{\partial\theta}\right)+\dfrac{1}{r^2}\dfrac{\partial^2\psi_1}{\partial\theta^2}-r\dfrac{\partial}{\partial r}\left(\dfrac{1}{r}\dfrac{\partial\psi_1}{\partial r}\right)\right]\end{cases} \tag{6.1.29}$$

The pile is considered as an elastic beam with a circular cross-sectional area A_p, Young's modulus E_p, mass density ρ_2, and Poisson's ratio υ_2. Thus the refracted fields inside the jth pile can be represented by

$$\varphi_{2s} = \sum_{m=0}^{\infty}J_m(p_{21}r_j)\left(A_{2m}^j\cos m\theta_j + B_{2m}^j\sin m\theta_j\right) \tag{6.1.30}$$

$$\psi_{2s} = \sum_{m=0}^{\infty}J_m(p_{23}r_j)\left(E_{2m}^j\sin m\theta_j + F_{2m}^j\cos m\theta_j\right) \tag{6.1.31}$$

where $p_{21}=\sqrt{\dfrac{(1-\upsilon_2)E_p}{(1+\upsilon_2)(1-2\upsilon_2)\rho_2}}$ and $p_{23}=\sqrt{\dfrac{E_p}{2(1+\upsilon_2)\rho_2}}$ are the wave numbers of longitudinal and transverse waves inside the piles, respectively; A_{2m}^j, B_{2m}^j, E_{2m}^j, and F_{2m}^j are unknown coefficients to be determined from boundary conditions.

The unknown scattering coefficients A_{1n}^j through F_{1n}^j in Eqs. (6.1.27), (6.1.28) must be determined by the application of suitable boundary conditions. It is assumed that during the vibrations no slippage or gap develops between the pile and the soil; namely, the pile is perfectly bonded to the soil, which is justified given the small amplitude of the ground-borne vibrations of interest. In addition, it is assumed that the interfaces are impermeable. Compatibility of displacements and stresses therefore requires

$$u_{1r}(r_j,\theta_j)\big|_{r_j=a_j}=u_{2r}(r_j,\theta_j)\big|_{r_j=a_j},\quad u_{1\theta}(r_j,\theta_j)\big|_{r_j=a_j}=u_{2\theta}(r_j,\theta_j)\big|_{r_j=a_j} \tag{6.1.32}$$

$$\sigma_{1rr}(r_j,\theta_j)\big|_{r_j=a_j}=\sigma_{2rr}(r_j,\theta_j)\big|_{r_j=a_j},\quad \tau_{1r\theta}(r_j,\theta_j)\big|_{r_j=a_j}=\tau_{2r\theta}(r_j,\theta_j)\big|_{r_j=a_j} \tag{6.1.33}$$

$$w_{1r}(r_j,\theta_j)\big|_{r_j=a_j}=0 \tag{6.1.34}$$

where θ_j ranges from 0 to 2π radians, j = 1, 2,…, N; a_j represents the radius of the jth pile; the subscripts "1" and "2" designate the soil and the pile, respectively.

Substituting Eqs. (6.1.27), (6.1.28) into Eq. (6.1.29), the displacements and stresses at each interface of the pile and soil can be obtained. Also, the boundary conditions at the pile-soil interfaces are considered. Then, taking into account the linear independence of trigonometric functions, 10 infinite linear systems of algebraic equations for A_{1m}^j through F_{2m}^j are obtained:

$$\begin{aligned}
&\frac{\varepsilon_m}{2}\sum_{n=0}^{\infty}\left[(1-\delta_{j1})\sum_{k=1}^{j-1}(-1)^m A_{1n}^k K_m^n(p_1 d_{kj})+(1-\delta_{jN})\sum_{k=j+1}^{N}(-1)^n A_{1n}^k K_m^n(p_1 d_{kj})\right]\cdot\Re_q^{11}\\
&+\frac{\varepsilon_m}{2}\sum_{n=0}^{\infty}\left[(1-\delta_{j1})\sum_{k=1}^{j-1}(-1)^m C_{1n}^k K_m^n(p_2 d_{kj})+(1-\delta_{jN})\sum_{k=j+1}^{N}(-1)^n C_{1n}^k K_m^n(p_2 d_{kj})\right]\cdot\Re_q^{12}\\
&+\frac{\varepsilon_m}{2}\sum_{n=0}^{\infty}\left[(1-\delta_{j1})\sum_{k=1}^{j-1}(-1)^m E_{1n}^k L_m^n(p_3 d_{kj})+(1-\delta_{jN})\sum_{k=j+1}^{N}(-1)^n E_{1n}^k L_m^n(p_3 d_{kj})\right]\cdot\Re_q^{13}\\
&+A_{1m}^j\cdot\Re_q^{14}+C_{1m}^j\cdot\Re_q^{15}+E_{1m}^j\cdot\Re_q^{16}+A_{2m}^j\cdot\Re_q^{17}+E_{2m}^j\cdot\Re_q^{18}\\
&=-\psi_0 e^{ip_3 d_{1j}\cos\theta_\beta}\varepsilon_m i^m\sin m\theta_\beta\cdot\Re_q^{13}
\end{aligned} \tag{6.1.35}$$

$$\begin{aligned}
&\frac{\varepsilon_m}{2}\sum_{n=0}^{\infty}\left[(1-\delta_{j1})\sum_{k=1}^{j-1}(-1)^m B_{1n}^k L_m^n(p_1 d_{kj})+(1-\delta_{jN})\sum_{k=j+1}^{N}(-1)^n B_{1n}^k L_m^n(p_1 d_{kj})\right]\cdot\Re_q^{21}\\
&+\frac{\varepsilon_m}{2}\sum_{n=0}^{\infty}\left[(1-\delta_{j1})\sum_{k=1}^{j-1}(-1)^m D_{1n}^k L_m^n(p_2 d_{kj})+(1-\delta_{jN})\sum_{k=j+1}^{N}(-1)^n D_{1n}^k L_m^n(p_2 d_{kj})\right]\cdot\Re_q^{22}\\
&+\frac{\varepsilon_m}{2}\sum_{n=0}^{\infty}\left[(1-\delta_{j1})\sum_{k=1}^{j-1}(-1)^m F_{1n}^k K_m^n(p_3 d_{kj})+(1-\delta_{jN})\sum_{k=j+1}^{N}(-1)^n F_{1n}^k K_m^n(p_3 d_{kj})\right]\cdot\Re_q^{23}\\
&+B_{1m}^j\cdot\Re_q^{24}+D_{1m}^j\cdot\Re_q^{25}+F_{1m}^j\cdot\Re_q^{26}+B_{2m}^j\cdot\Re_q^{27}+F_{2m}^j\cdot\Re_q^{28}\\
&=-\psi_0 e^{ip_3 d_{1j}\cos\theta_\beta}\varepsilon_m i^m\cos m\theta_\beta\cdot\Re_q^{23}
\end{aligned} \tag{6.1.36}$$

where q = 1, 2,…, 5; j = 1, 2, · · ·, N; m = 0, 1, 2,…, ∞. R_q^{11} through R_q^{28} are given in Appendix G.

Once the systems of Eqs. (6.1.35), (6.1.36) are truncated and solved with an appropriate range for the expansions, the scattering wave field and the total wave field can be obtained, and the corresponding displacements and stresses in each point of the poroelastic soil can be subsequently calculated.

Incidence of Plane Fast Compressional Waves

Similarly, the potential function for an incidence of a fast compressional wave with amplitude φ_0 is represented by a Bessel series:

$$\varphi^{(\mathrm{inc})}(r_1, \theta_1) = \varphi_0 \mathrm{e}^{p_1 d_{1j} \cos\theta_\alpha} \cdot \sum_{m=0}^{\infty} \varepsilon_m \mathrm{i}^m J_m(p_1 r_j) \cos\left[m(\theta_j - \theta_\alpha)\right] \tag{6.1.37}$$

where θ_α is the incident angle.

It is noted that the scattered waves in the solid matrix are the same as in Eq. (6.1.20), and the refracted fields inside the jth pile are the same as in Eqs. (6.1.30), (6.1.31). Thus for an incident fast compressional wave, the right sides of Eqs. (6.1.30), (6.1.31) should be replaced by Eqs. (6.1.38), (6.1.39), respectively.

$$-\varphi_0 \mathrm{e}^{\mathrm{i}p_1 d_{1j} \cos\theta_\alpha} \varepsilon_m \mathrm{i}^m \cos m\theta_\alpha \cdot \Re_q^{11} \tag{6.1.38}$$

$$-\varphi_0 \mathrm{e}^{\mathrm{i}p_1 d_{1j} \cos\theta_\alpha} \varepsilon_m \mathrm{i}^m \sin m\theta_\alpha \cdot \Re_q^{21} \tag{6.1.39}$$

6.1.3 Numerical Results and Conclusions

For numerical calculations of the soil displacement distributions behind a row of elastic piles, a truncation to a finite size for the systems of Eqs. (6.1.35), (6.1.36) is made. By performing computations over a wide range of complex arguments and integer orders, it is found that, to provide an accuracy of three significant figures for the frequencies studied, the order of the expansions should be truncated to 6.

When a time-harmonic plane wave of unit amplitude propagates perpendicularly to a row of piles, the wave is scattered into a combination of fast and slow compressional waves and transverse waves. In addition, for the convenience of numerical calculations, the piles are assumed to be identical and equally spaced with a radius a. Some notations were used in the analysis: Young's modulus = E_p, mass density = ρ_2, and pile spacing sp, denoting the distance between the centers of adjacent piles (Fig. 6.1.2). The length of the discontinuous barrier of piles is L, and the origin of the rectangular coordinate system is placed at the center of the barrier as illustrated in Fig. 6.1.2. To facilitate the presentation of the numerical results, the dimensionless material properties and frequency are introduced by Eqs. (4.1.10), (6.1.40) is as follows:

$$m^* = \frac{h}{\rho}, \quad \omega^* = \frac{\omega d}{C_\mathrm{s}}, \quad \rho_2^* = \frac{\rho_2}{\rho}, \quad E_\mathrm{p}^* = \frac{E_\mathrm{p}}{E_\mathrm{s}} \tag{6.1.40}$$

where $d = 2a$ is the diameter of the pile, C_s denotes the transverse wave velocity in poroelastic soil, and E_s represents Young's modulus of the skeletal frame.

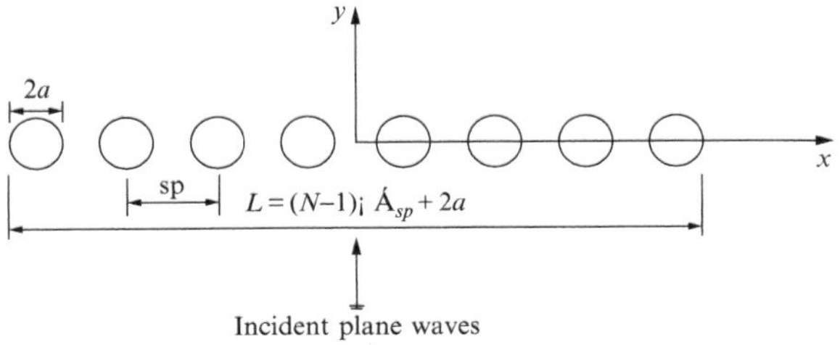

Fig. 6.1.2 A row of piles and a rectangular coordinate system

The normalized displacement amplitudes $|u_x/u_0|$ and $|v_y/v_0|$ are defined as the ratios of the displacement amplitude of the soil in the presence of piles over that displacement in the absence of piles along the x and y axis, respectively, and y/a denotes the dimensionless distance away from the barrier.

Comparisons with Existing Solutions for Ideal Elastic Soil

It is noted that if the poroelastic parameters M^*, ρ^*, m^*, b^*, and α are assumed to be nil, the poroelastic medium is reduced to an elastic one. To verify the proposed method for calculation of the amplitude reduction of elastic waves by a solid pile barrier, our results compare with those of Avilés and Sánchez-Sesma[44]. The following parameters for soil and pile are used in the calculation: ρ_2/ρ = 1.54, sp/a = 3.0, y/a = 125, N = 8, and E_p/E_s = 10,000. The normalized frequencies of the incident plane longitudinal and transverse waves are taken as 0.45 and 0.8, respectively. Fig. 6.1.3 shows a comparison of the normalized displacement amplitudes between the results obtained from the present study and that of Avilés and Sánchez-Sesma[44]. A fairly good agreement is found between them, such that the accuracy of the present solution scheme is validated.

In the following, some numerical examples are presented to investigate the influence of certain parameters on the screening of plane waves achieved by piles in a row. The properties of the soil and piles are summarized in Table 6.1.1.

Isolation of Incident Transverse Waves by a Row of Elastic Piles

The influence of internal friction (b^*) resulting from relative motion between the solid skeleton and the pore fluid is investigated in Figs. 6.1.4 and 6.1.5, by varying values of b^* (b^* = 0, 5, 50, and 500). Recalling b^* is inversely proportional to the permeability, it suggests that the material with b^* = 0 is the most permeable, and the material with b^* = 500 is the least permeable, among the four poroelastic materials. The material property b^* is found to have a more significant influence on the isolation effectiveness when compared to other parameters of the poroelastic material. Figs. 6.1.4 and 6.1.5 clearly show that there is a gradual improvement of isolation effectiveness as b^* is increased to a certain value. This phenomenon may be due to the coupling effect of the fluid and the soil skeleton, which is quite significant if the soil has a low intrinsic permeability. However, it is seen that the b^* value around 500 appears to be a threshold, beyond which an increase in b^* could hardly promote the screening efficiency.

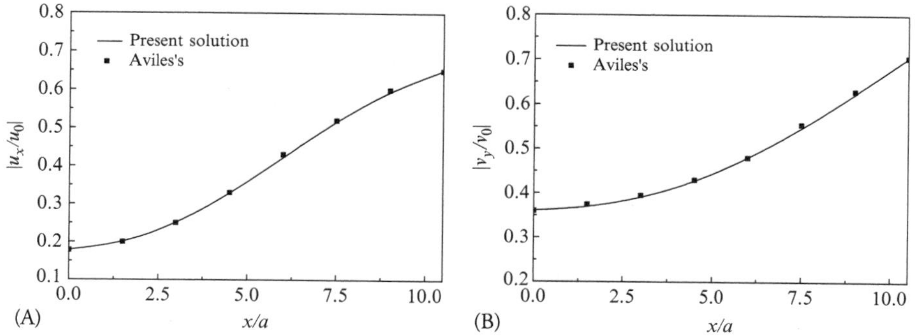

Fig. 6.1.3 Comparisons of normalized displacement amplitudes between the present solutions and those of Avilés and Sánchez-Sesma for elastic soil. (A) Transverse plane wave incidence and (B) longitudinal plane wave incidence

Table 6.1.1 Dimensionless parameters used in the calculation

Parameter	Value
α	1.00
λ^*	1.004
M^*	246.78
ρ_f^*	0.45
m^*	1.68
ρ_2^*	1.13
b^*	50
υ_2	0.2
E_p/E_s	100

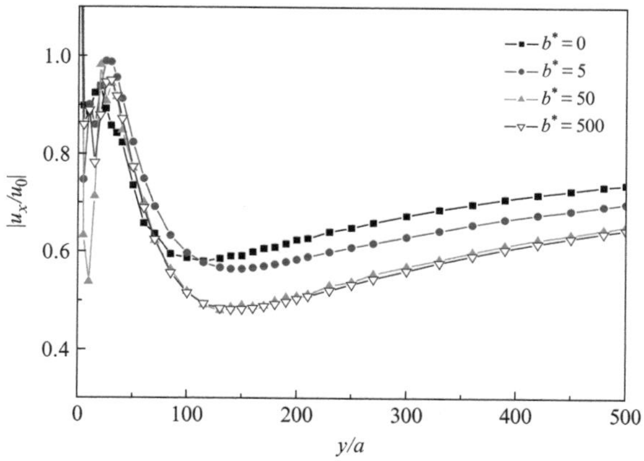

Fig. 6.1.4 $|u_x/u_0|$ versus y/a with different values of b^* for $\omega^* = 1.2$, $N = 8$, $sp/a = 3$, $E_p/E_s = 100$, and $x/a = 0$

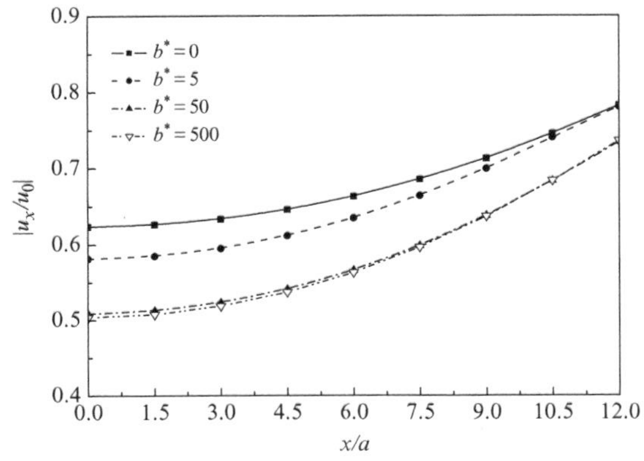

Fig. 6.1.5 $|u_x/u_0|$ versus x/a with different values of b^* for $\omega^*=1.2$, $N=8$, $sp/a=3$, $E_p/E_s=100$, and $y/a=200$

The pile spacing is another key factor that directly affects the isolation effectiveness of the barrier in poroelastic soil. In this study, four different values of the pile-spacing ratio are considered: sp/a = 2.5, 3.0, 3.5, and 4.0. Fig. 6.1.6 suggests that the minimum value of $|u_x/u_0|$ grows as the pile spacing is increased, meaning that an increase in pile spacing may result in a poorer isolation effectiveness. This observation highlights the principal role that the pile spacing plays in the screening of elastic waves. Furthermore, at points near the row of piles, the screened zone shows a large variability in contrast to the smooth variations observed at the locations far away from the barrier. A comparison between Figs. 6.1.6A and B indicates that the normalized displacement amplitude $|u_x/u_0|$ is larger at the edge (x/a = 10.5) than at the center (x/a = 0), with the same pile spacing, suggesting that the isolation effectiveness in the center of the screened zone is much better than that at the edge. The possible reason is given as follows: when the elastic wave propagates through the pile barrier, an interference effect exists due to the wave scattering. The scattering waves are stronger in the center of the pile barrier than at the edge, where more wave energy is transmitted to the screened zone because of the finite size of the barrier. Fig. 6.1.6B also

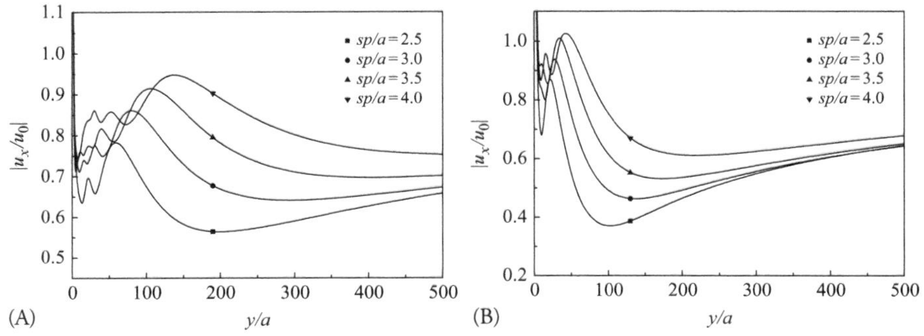

Fig. 6.1.6 $|u_x/u_0|$ versus y/a with different values of pile spacing for $\omega^*=1.2$, $N=8$, $E_p/E_s=100$, and $b^*=50$. (A) at the edge of the screened zone; (B) at the center of the screened zone

demonstrates that the wave field presents a uniform recovery and tends to approach the levels without the row of piles in the width observed far away from the barrier. Moreover, as expected, the piles should be very close to one another for the sake of obtaining an optimal isolation effectiveness.

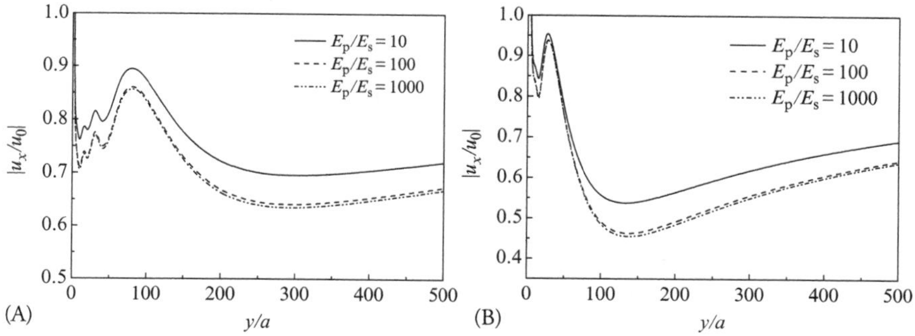

Fig. 6.1.7 $|u_x/u_0|$ versus y/a with different Young's modulus of piles for ω^*=1.2, N=8, sp/a=3, and b^*=50. (A) at the edge of the screened zone; (B) at the center of the screened zone

The influence of Young's modulus of the piles on the isolation effectiveness is illustrated in Figs. 6.1.7 and 6.1.8. It can be seen from these figures that the optimal screening position is in the center of the screened zone. It is also seen in Fig. 6.1.7 that an increase in Young's modulus of piles results in improved isolation effectiveness of the pile barrier for transverse plane waves. In the center of the screened zone, as E_p/E_s is varied from 10 to 100, the minimum value of $|u_x/u_0|$ changes from 0.54 to 0.46. That is, the isolation effectiveness of the barrier has an increase of 8%. However, as depicted in Figs. 6.1.7 and 6.1.8, when Young's modulus of piles goes up to a certain extent (E_p/E_s > 100), the further increase of Young's modulus of piles appears to have little influence on the isolation effectiveness. Hence, it is recommended that E_p/E_s values should be larger than 10 and preferably at least 100. E_p/E_s values larger than 1000 are not considered, because they produce only a slight improvement in the screening efficiency.

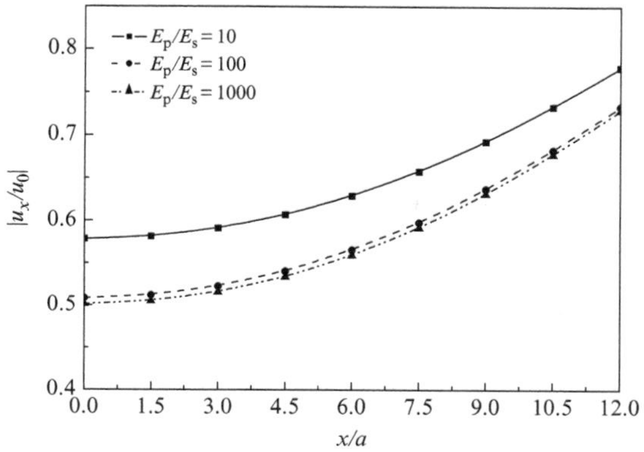

Fig. 6.1.8 $|u_x/u_0|$ versus x/a with a different Young's modulus of piles for ω^*=1.2, N=8, sp/a=3, b^*=50, and y/a = 200

As shown before, most previous investigations of the present problem are limited to the elastic medium. Because the elastic medium model cannot predict the influence of the permeability of the soil on the dynamic response of the piles, the present model is preferential to poroelastic soil. In this regard, it is of interest here to compare the isolation effectiveness of the pile barrier in poroelastic soil with that under elastic conditions.

Fig. 6.1.9 is a comparison of normalized amplitudes of displacements, u_x, versus y/a between elastic and poroelastic soils with three values of the dimensionless frequency (ω*=0.6, 1.2, and 1.8) for b*=0 and 500, respectively. The differences between elastic and poroelastic soils are primarily in the magnitude of $|u_x/u_0|$, rather than in their general behavior. It seems that the isolation effectiveness improves remarkably with an increase of ω*. The displacement amplitude is increased or decreased due to the phase mismatch between the incident and scattering waves after the superposition, which leads to the appearance of random values among all curves for $y/a < 50$. As shown in Fig. 6.1.9, there is much resemblance between poroelastic material with b*=0 and elastic soil for the isolation effectiveness of a pile barrier, while the isolation effectiveness in poroelastic soil with b*=500 is better than that of the elastic condition when $y/a > 150$. Another important feature of Fig. 6.1.9B is that the differences in $|u_x/u_0|$ between poroelastic and elastic soils become more marked as the dimensionless frequency ω* increased. These results clearly indicate that the elastic solutions can hardly be used to evaluate the screening performance of the pile barrier in poroelastic soil except for smaller values of b*.

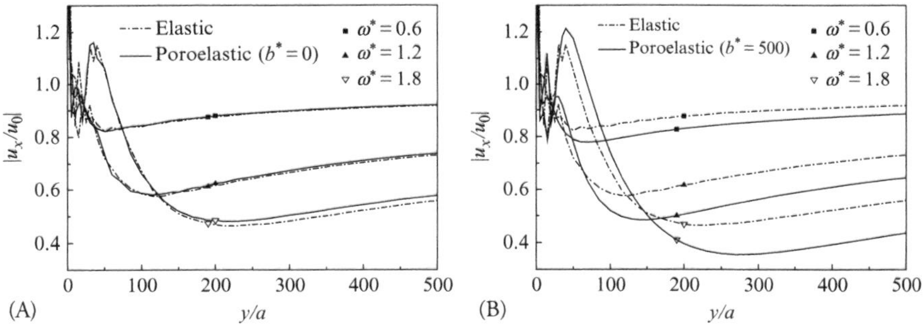

Fig. 6.1.9 Comparisons of $|u_x/u_0|$ between poroelastic soil and elastic soil under different dimensionless frequencies for $N = 8$, $sp/a = 3$, $E_p/E_s = 100$, and $x/a = 0$. (A) $b^* = 0$ and (B) $b^* = 500$

A comparison of $|u_x/u_0|$ versus y/a between elastic and poroelastic soils with three values of the pile spacing ratio (sp/a = 2.5, 3.0, and 3.5) is presented in Fig. 6.1.10. A remarkable difference between the curves for elastic soil and poroelastic soil with b*=500 is observed, meaning simply that the isolation effectiveness in poroelastic soil with b*=500 is better than that under the elastic condition when $y/a > 80$. The behavior of $|u_x/u_0|$ in elastic soil almost follows the pattern of poroelastic soil of b*=0 with the same pile spacing. All these results suggest that, in the case of sand in which b* is very small, the shape of the curves is similar to that of the elastic condition, which shows a slight difference in the normalized displacement amplitude $|u_x/u_0|$. This is due to the fact that, for a small value of b*, the coupling effect produced by motion of the solid matrix and the pore fluid is limited. However, in the case of soft clay, in which b* is much larger, a significant difference in $|u_x/u_0|$ can be observed between poroelastic soil and elastic soil, and it might be attributed to the presence of the pore fluid and the effect of its dissipation from the pores of the solid matrix.

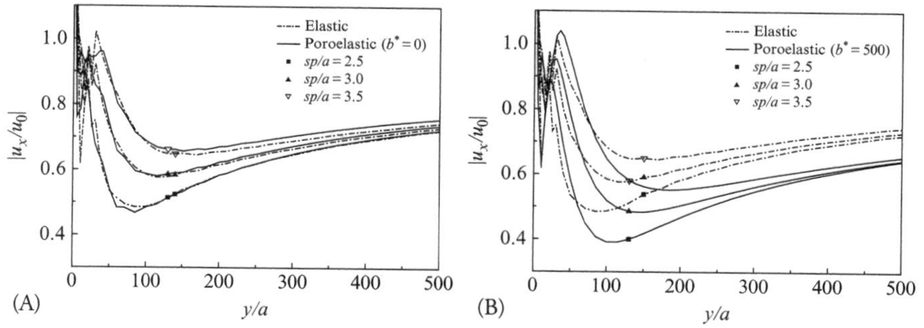

Fig. 6.1.10 Comparisons of $|u_x/u_0|$ between poroelastic soil and elastic soil under different values of pile spacing for ω^* = 1.2, N = 8, E_p/E_s = 100, and x/a = 0. (A) b^* = 0 and (B) b^* = 500

The influence of the number of piles on the screening efficiency in elastic and poroelastic soil is shown in Fig. 6.1.11, where the number of piles is varied from 6 to 10. As the number of piles is increased, the optimal screening position behind the row of piles is moved backwards, and the normalized displacement amplitude $|u_x/u_0|$ decreases when y/a > 170. The curves also show the same trends in elastic soil and in poroelastic soil with b^* = 0, while in poroelastic soil with b^* = 500, there is a sharp decrease in $|u_x/u_0|$ in comparison with that under the elastic condition when y/a > 170. The physical explanation of this phenomenon is that there is a viscous resistance for a flowing fluid in poroelastic soil with b^* = 500, which gives rise to the energy dissipation from the pores of the solid matrix.

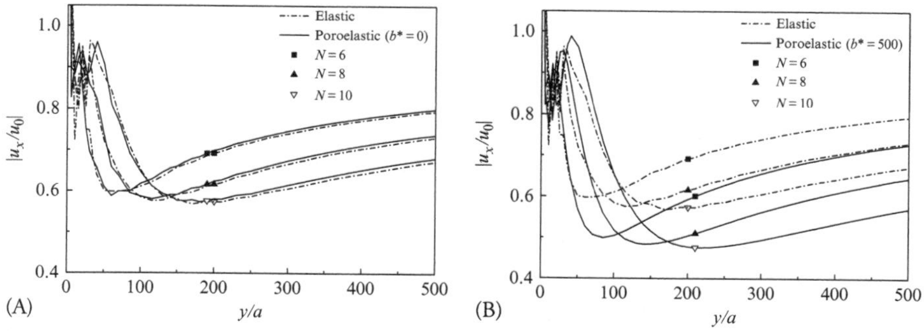

Fig. 6.1.11 Comparisons of $|u_x/u_0|$ between poroelastic soil and elastic soil under different numbers of piles for ω^*=1.2, sp/a=3, E_p/E_s=100, and x/a=0. (A) b^* =0 and (B) b^*=500

The influence of Young's modulus on the screening efficiency in elastic and poroelastic soils is illustrated in Fig. 6.1.12. The numerical results show that as Young's modulus of piles is increased, the normalized displacement amplitude $|u_x/u_0|$ decreases. Nevertheless, for E_p/E_s > 100, the decrease in $|u_x/u_0|$ is quite small, due to an increase in the ratio of Young's moduli of the pile and the soil. The curves also show the same trends in elastic soil and in poroelastic soil with b^* = 0, while in poroelastic soil with b^* = 500, there is a dramatic decrease in $|u_x/u_0|$ as compared with that in elastic soil when y/a > 120. Furthermore, the optimal isolation effectiveness in poroelastic soil with b^*= 500 is about 10% higher than that under the elastic condition when $E_p/E_s = 100$.

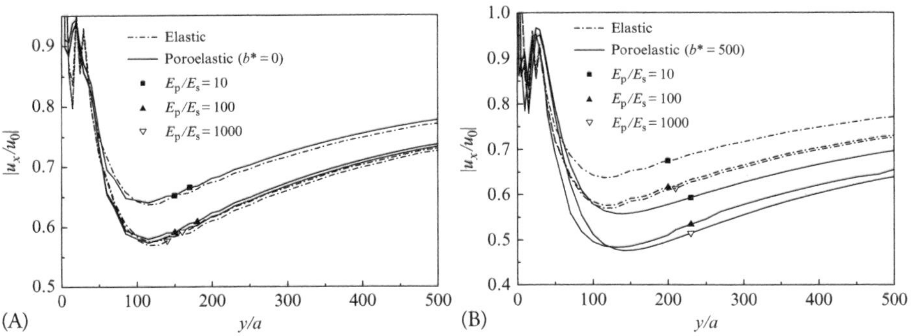

Fig. 6.1.12 Comparisons of $|u_x/u_0|$ between poroelastic soil and elastic soil under different values of Young's modulus for $\omega^* = 1.2$, $sp/a = 3$, $N = 8$, and $x/a = 0$. (A) $b^* = 0$ and (B) $b^* = 500$

Aside from the phenomena observed so far, Figs. 6.1.9–6.1.12 also show an unusual trend for the case of $y/a < 50$. The displacement amplitude of soil is increased or decreased after the superposition due to the interaction of incident and scattering waves. It is also noted from Figs. 6.1.4–6.1.12 that the curves go up and not down. That is to say, all of the curves for the poroelastic model come together and tend to reach the levels without the barrier in the limit of large y/a. This may be due to the fact that the scattering wave field at a larger distance becomes weaker, which leads to a weak interaction between the incident wave field and that of the scattering wave.

6.1.4 Conclusions

The focus of the current work was to investigate the effects in isolation of elastic waves achieved by a row of piles in a poroelastic medium. A set of cylindrical coordinate systems and Graff's theorem for cylindrical Bessel functions were used to present analytical solutions as well as numerical results for the dynamic interaction of piles in a boundless poroelastic medium, obeying Biot's dynamic model. Based on the results of the parametric study, the following points can be summarized:

(1) The isolation effectiveness in clay (high b^* value) becomes more significant compared to that in sand (low b^* value) for viscous losses. Therefore, for poroelastic soil with a small coefficient of permeability, such as soft clay, it is preferable to use the poroelastic medium model to achieve more reasonable and reliable results.

(2) The screening performance of a pile barrier for transverse plane waves improves as Young's modulus of piles increases. However, after an E_p/E_s value of around 100, the effectiveness of the improvement is rather limited.

(3) The pile spacing is one of the key parameters that affect the screening efficiency of a pile barrier. Generally, the larger the pile spacing, the poorer the isolation effectiveness of a pile barrier would be in poroelastic soil in the case of the incidence of transverse waves. In addition, the isolation effectiveness of a pile barrier improves as the number of piles increases in poroelastic soil, and the isolation effectiveness in the center of the screened zone is better than that at the edge.

(4) The row of piles serving as a whole barrier for isolation of fast incident compressional waves is not apparent when the excitation frequency is low. To avoid this problem, multiple rows of piles can be employed as an alternative for the vibration isolation in poroelastic soil.

6.2 Isolation of Rayleigh Waves in Poroelastic Soil

6.2.1 Solution of Governing Equations for a Poroelastic Medium

The equations of the motion of a poroelastic medium in terms of displacements are given in Eqs. (1.1.24), (1.1.25).

We selected the rectangular Cartesian coordinates (x, y, z) with the z = 0 plane being the boundary of the poroelastic half-space with the positive z axis lying outside the medium. The displacements of the solid skeleton and pore fluid in the y and z directions, denoted respectively by u_y, u_z and w_y, w_z, can be represented in terms of potentials as follows:

$$\begin{cases} u_y = \dfrac{\partial \varphi_s}{\partial y} - \dfrac{\partial \psi_s}{\partial z}, \quad u_z = \dfrac{\partial \varphi_s}{\partial z} + \dfrac{\partial \psi_s}{\partial y} \\ w_y = \dfrac{\partial \varphi_f}{\partial y} - \dfrac{\partial \psi_f}{\partial z}, \quad w_z = \dfrac{\partial \varphi_f}{\partial z} + \dfrac{\partial \psi_f}{\partial y} \end{cases} \tag{6.2.1}$$

where φ_s and ψ_s are potentials associated with the solid phase of the bulk material, while potentials φ_f and ψ_f are associated with the pore fluid phase.

Substituting Eq. (6.2.1) into Eqs. (1.1.24), (1.1.25), with some manipulations, the four potentials that represent plane harmonic waves traveling in the y direction can be expressed as

$$\begin{cases} \varphi_s = \left(A_1 e^{ka_1 z} + A_2 e^{ka_2 z}\right) \cdot e^{-ik(y-ct)} \\ \varphi_f = \left(A_1 B_1 e^{ka_1 z} + A_2 B_2 e^{ka_2 z}\right) \cdot e^{-ik(y-ct)} \\ \psi_s = A_3 e^{kb_1 z} \cdot e^{-ik(y-ct)} \\ \psi_f = \dfrac{ib}{ib - \rho_2 \omega} A_3 e^{kb_1 z} \cdot e^{-ik(y-ct)} \end{cases} \tag{6.2.2}$$

where A_1, A_2, and A_3 are constants; $i=\sqrt{-1}$; $k=\omega/c$ is the wave number, in which c is the Rayleigh speed and ω the frequency; $a_1=1-c^2/V_{p1}^2$, $a_2=1-c^2/V_{p2}^2$, V_{p1} and V_{p2} denote the velocities of the dilatational wave of the first and second kind; $b_1 = 1 - c^2/V_s^2$, $\dfrac{1}{V_s^2} = \dfrac{\rho_1}{\mu} + \dfrac{ib\rho_2}{\mu(ib-\rho_2\omega)}$;

$$B_h = \frac{(\lambda+2\mu)\omega}{ib}\left(\frac{1}{V_{ph}^2} - \frac{\rho_1}{\lambda+2\mu}\right) + 1, \quad h=1,2.$$

The relevant solid stresses σ_y, σ_z, and τ_{yz}, along with liquid pressure p_f, can be written in terms of the potentials as

$$\begin{cases} \sigma_y = \lambda\left(\dfrac{\partial^2\varphi_s}{\partial y^2} + \dfrac{\partial^2\varphi_s}{\partial z^2}\right) + 2\mu\left(\dfrac{\partial^2\varphi_s}{\partial y^2} - \dfrac{\partial^2\psi_s}{\partial y\partial z}\right) \\ \sigma_z = \lambda\left(\dfrac{\partial^2\varphi_s}{\partial y^2} + \dfrac{\partial^2\varphi_s}{\partial z^2}\right) + 2\mu\left(\dfrac{\partial^2\varphi_s}{\partial z^2} + \dfrac{\partial^2\psi_s}{\partial y\partial z}\right) \\ \tau_{yz} = \mu\left(2\dfrac{\partial^2\varphi_s}{\partial y\partial z} + \dfrac{\partial^2\psi_s}{\partial y^2} - \dfrac{\partial^2\psi_s}{\partial z^2}\right) \\ p_f = \rho_2\ddot{\varphi}_f + b(\dot{\varphi}_f - \dot{\varphi}_s) \end{cases} \tag{6.2.3}$$

It was assumed that the surface of the poroelastic half-space is pervious, so the boundary conditions at $z = 0$ can be obtained as

$$\sigma_z + p_f = 0,\ \tau_{yz} = 0,\ p_f = 0 \tag{6.2.4}$$

Substituting Eq. (6.2.3) into Eq. (6.2.4) yields

$$\begin{cases} \sigma_z = \left[(\lambda+2\mu)a_1^2 - \lambda\right]k^2A_1 + \left[(\lambda+2\mu)a_2^2 - \lambda\right]k^2A_2 - 2\mathrm{i}\mu b_1k^2A_3 = 0 \\ \tau_{yz} = -2\mathrm{i}\mu a_1k^2A_1 - 2\mathrm{i}\mu a_2k^2A_2 - \mu\left(b_1^2+1\right)k^2A_3 = 0 \\ p_f = \left[\mathrm{i}b\omega(B_1-1) - B_1\rho_2\omega^2\right]A_1 + \left[\mathrm{i}b\omega(B_2-1) - B_2\rho_2\omega^2\right]A_2 = 0 \end{cases} \tag{6.2.5}$$

The conditions for the nontriviality of the solution for A_1, A_2, and A_3 are the vanishing of the determinant of the coefficients in Eq. (6.2.5). Thus the Rayleigh wave speed equations for the pervious surface can be obtained via the following form:

$$\begin{vmatrix} (\lambda+2\mu)a_1^2 - \lambda & (\lambda+2\mu)a_2^2 - \lambda & -2\mathrm{i}\mu b_1 \\ 2\mathrm{i}a_1 & 2\mathrm{i}a_2 & 1+b_1^2 \\ \mathrm{i}b(B_1-1) - B_1\rho_2\omega & \mathrm{i}b(B_2-1) - B_2\rho_2\omega & 0 \end{vmatrix} = 0 \tag{6.2.6}$$

6.2.2 Isolation of Rayleigh Waves by a Row of Piles

Consider Rayleigh waves propagating toward a row of piles as shown in Fig. 6.2.1. Such incident waves are given in the reference system (r_l, θ_l, z) attached to the lth pile by means of

$$\begin{cases} \varphi_s^{\mathrm{inc}} = \left(A_1\mathrm{e}^{ka_1z} + A_2\mathrm{e}^{ka_2z}\right)\cdot\mathrm{e}^{-\mathrm{i}kr_l\sin\theta_l}\cdot\mathrm{e}^{\mathrm{i}\omega t} \\ \varphi_f^{\mathrm{inc}} = \left(A_1B_1\mathrm{e}^{ka_1z} + A_2B_2\mathrm{e}^{ka_2z}\right)\cdot\mathrm{e}^{-\mathrm{i}kr_l\sin\theta_l}\cdot\mathrm{e}^{\mathrm{i}\omega t} \\ \psi_s^{\mathrm{inc}} = A_3\mathrm{e}^{kb_1z}\cdot\mathrm{e}^{-\mathrm{i}kr_l\sin\theta_l}\cdot\mathrm{e}^{\mathrm{i}\omega t} \\ \psi_f^{\mathrm{inc}} = \dfrac{\mathrm{i}b}{\mathrm{i}b - \rho_2\omega}A_3\mathrm{e}^{kb_1z}\cdot\mathrm{e}^{-\mathrm{i}kr_l\sin\theta_l}\cdot\mathrm{e}^{\mathrm{i}\omega t} \end{cases} \tag{6.2.7}$$

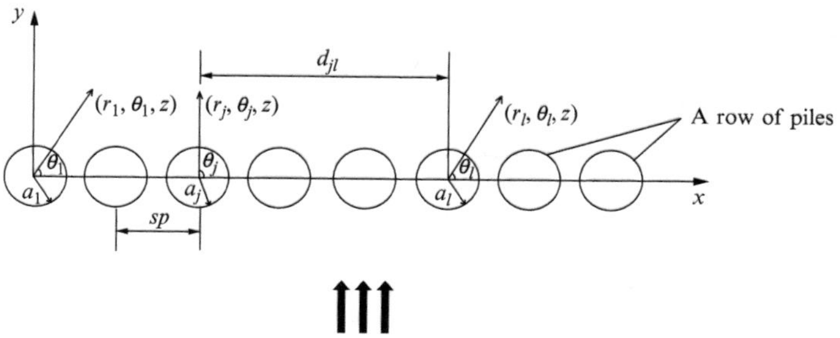

Fig. 6.2.1 Incidence of Rayleigh waves and reference systems

The present analysis pertains to steady-state harmonic excitation with frequency ω. Because the factor $e^{i\omega t}$ is shared by all the variables (excitations and responses), it will be omitted from now on for simplicity.

Expanding the factor $e^{-ikr_l \sin\theta_l}$ in the Fourier-Bessel series yields

$$e^{-ikr_l \sin\theta_l} = \sum_{m=0}^{\infty} \varepsilon_m (-i)^m J_m(kr_l) \cos\left[m\left(\theta_l - \frac{\pi}{2}\right)\right] \tag{6.2.8}$$

where $\varepsilon_0=1$, $\varepsilon_m=2$ when $m\geq 1$, and $J_m(\cdot)$ is the Bessel function of the first kind and order m.

The scattered field in the poroelastic soil by the jth pile can be expressed in the form

$$\begin{cases} \varphi_s^{sc} = \left(A_1 e^{ka_1 z} + A_2 e^{ka_2 z}\right) \cdot \sum_{n=0}^{\infty} \left(A_n^j \cos n\theta_j + B_n^j \sin n\theta_j\right) H_n^{(2)}(kr_j) \\ \varphi_f^{sc} = \left(A_1 B_1 e^{ka_1 z} + A_2 B_2 e^{ka_2 z}\right) \cdot \sum_{n=0}^{\infty} \left(A_n^j \cos n\theta_j + B_n^j \sin n\theta_j\right) H_n^{(2)}(kr_j) \\ \psi_s^{sc} = A_3 e^{kb_1 z} \cdot \sum_{n=0}^{\infty} \left(A_n^j \cos n\theta_j + B_n^j \sin n\theta_j\right) H_n^{(2)}(kr_j) \\ \psi_f^{sc} = \dfrac{ib}{ib - \rho_2 \omega} A_3 e^{kb_1 z} \cdot \sum_{n=0}^{\infty} \left(A_n^j \cos n\theta_j + B_n^j \sin n\theta_j\right) H_n^{(2)}(kr_j) \end{cases} \tag{6.2.9}$$

where φ_s^{sc}, φ_f^{sc}, ψ_s^{sc}, and ψ_f^{sc} represent the waves scattered by the jth pile in terms of the jth system of cylindrical coordinates (r_j, θ_j, z), $H_n^{(2)}(\cdot)$ is the Hankel function of the second kind and order n, which represents an outgoing wave that satisfies the Sommerfeld radiation condition at infinity, and A_n^j, B_n^j are complex coefficients to be determined from boundary conditions.

In the presence of a row of piles, the incident Rayleigh waves will be scattered and diffracted around the piles. The total wave fields near the jth pile are therefore given by

$$\begin{cases} \varphi_{\mathrm{s}} = \varphi_{\mathrm{s}}^{\mathrm{inc}} + \sum_{j=1}^{N} \varphi_{\mathrm{s}}^{\mathrm{sc}}, & \psi_{\mathrm{s}} = \psi_{\mathrm{s}}^{\mathrm{inc}} + \sum_{j=1}^{N} \psi_{\mathrm{s}}^{\mathrm{sc}} \\ \varphi_{\mathrm{f}} = \varphi_{\mathrm{f}}^{\mathrm{inc}} + \sum_{j=1}^{N} \varphi_{\mathrm{f}}^{\mathrm{sc}}, & \psi_{\mathrm{f}} = \psi_{\mathrm{f}}^{\mathrm{inc}} + \sum_{j=1}^{N} \psi_{\mathrm{f}}^{\mathrm{sc}} \end{cases} \tag{6.2.10}$$

where N is the number of piles.

The wave potentials with respect to the cylindrical coordinate systems (r_j, θ_j, z) and (r_l, θ_l, z), in which $j \neq l$, are related through the following transformation:

$$\begin{aligned} & H_n^{(2)}(kr_j)\left(A_n^j \cos n\theta_j + B_n^j \sin n\theta_j\right) \\ & = (1-\delta_{l1})\frac{1}{2}\sum_{j=1}^{l-1}\sum_{m=0}^{\infty}(-1)^m \varepsilon_m J_m(kr_l)\left[A_n^j K_m^n(kd_{jl})\cos m\theta_l + B_n^j L_m^n(kd_{jl})\sin m\theta_l\right] \\ & + H_n^{(2)}(kr_l)\left(A_n^l \cos n\theta_l + B_n^l \sin n\theta_l\right) \\ & + (1-\delta_{lN})\frac{1}{2}\sum_{j=l+1}^{N}\sum_{m=0}^{\infty}(-1)^n \varepsilon_m J_m(kr_l)\left[A_n^j K_m^n(kd_{jl})\cos m\theta_l + B_n^j L_m^n(kd_{jl})\sin m\theta_l\right] \end{aligned} \tag{6.2.11}$$

where d_{jl} denotes the distance between the jth pile and the lth pile;

$$K_m^n(kd_{jl}) = H_{m+n}^{(2)}(kd_{jl}) + (-1)^n H_{m-n}^{(2)}(kd_{jl}) \tag{6.2.12}$$

$$L_m^n(kd_{jl}) = -H_{m+n}^{(2)}(kd_{jl}) + (-1)^n H_{m-n}^{(2)}(kd_{jl}) \tag{6.2.13}$$

In the cylindrical coordinates (r_j, θ_j, z) the displacement and stress components of the skeletal frame as well as the pore water pressure, by using the scalar potentials φ_{s}, ψ_{s}, φ_{f}, and ψ_{f}, can be derived as

$$u_r = \frac{\partial \varphi_{\mathrm{s}}}{\partial r} + \frac{\partial^2 \psi_{\mathrm{s}}}{\partial r \partial z} \tag{6.2.14}$$

$$u_\theta = \frac{1}{r}\frac{\partial \varphi_{\mathrm{s}}}{\partial \theta} + \frac{1}{r}\frac{\partial^2 \psi_{\mathrm{s}}}{\partial \theta \partial z} \tag{6.2.15}$$

$$u_z = \frac{\partial \varphi_{\mathrm{s}}}{\partial z} - \left[\frac{1}{r}\frac{\partial}{\partial r}\left(r\frac{\partial \psi_{\mathrm{s}}}{\partial r}\right) + \frac{1}{r^2}\frac{\partial^2 \psi_{\mathrm{s}}}{\partial \theta^2}\right] \tag{6.2.16}$$

$$\sigma_{rr} = \lambda \nabla^2 \varphi_{\mathrm{s}} + 2\mu\left[\frac{\partial^2 \varphi_{\mathrm{s}}}{\partial r^2} + \frac{\partial^3 \psi_{\mathrm{s}}}{\partial r^2 \partial z}\right] \tag{6.2.17}$$

$$\sigma_{r\theta} = 2\mu\left[\left(\frac{1}{r}\frac{\partial^2 \varphi_{\mathrm{s}}}{\partial r \partial \theta} - \frac{1}{r^2}\frac{\partial \varphi_{\mathrm{s}}}{\partial \theta}\right) + \left(\frac{1}{r}\frac{\partial^3 \psi_{\mathrm{s}}}{\partial r \partial \theta \partial z} - \frac{1}{r^2}\frac{\partial^2 \psi_{\mathrm{s}}}{\partial \theta \partial z}\right)\right] \tag{6.2.18}$$

$$\sigma_{rz} = \mu\left[2\frac{\partial^2 \varphi_{\mathrm{s}}}{\partial r \partial z} + \left(2\frac{\partial^3 \psi_{\mathrm{s}}}{\partial r \partial z^2} - \frac{\partial}{\partial r}\nabla^2 \psi_{\mathrm{s}}\right)\right] \tag{6.2.19}$$

$$p_f = -\rho_2\omega^2\varphi_f + b\omega \mathrm{i}(\varphi_f - \varphi_s) \tag{6.2.20}$$

where $\nabla^2 = \dfrac{\partial^2}{\partial r^2} + \dfrac{1}{r}\dfrac{\partial}{\partial r} + \dfrac{1}{r^2}\dfrac{\partial^2}{\partial \theta^2} + \dfrac{\partial^2}{\partial z^2}$.

The pile was modeled as an Euler-Bernoulli beam, embedded in a poroelastic half-space where Rayleigh waves were traveling. In addition, it was assumed that the interfaces are impermeable. In general, the bending rigidity of the piles tends to resist the induced deflections and the axial rigidity of the piles tends to resist the induced elongations/contractions. Accordingly, dynamic equilibrium under steady-state motion of the piles gives

$$E_pI_p\frac{\mathrm{d}^4U_l(z)}{\mathrm{d}z^4} - \omega^2 m_p U_l(z) = p_x^l(z) \tag{6.2.21}$$

$$E_pI_p\frac{\mathrm{d}^4V_l(z)}{\mathrm{d}z^4} - \omega^2 m_p V_l(z) = p_y^l(z) \tag{6.2.22}$$

$$E_pA_p\frac{\mathrm{d}^2W_l(z)}{\mathrm{d}z^2} + \omega^2 m_p W_l(z) = -p_z^l(z) \tag{6.2.23}$$

where $U_l(z)$, $V_l(z)$, and $W_l(z)$ are complex amplitudes of the motions of the lth pile in the x, y, and z directions, respectively; $-H \le z \le 0$, H is the length of the piles; $l = 1, 2, \ldots, N$; E_pI_p is the bending rigidity of the piles; E_pA_p is the axial rigidity of the piles; m_p is the mass of the piles per unit length; $p_x^l(z)$, $p_y^l(z)$, and $p_z^l(z)$ are amplitudes of the forces generated by the poroelastic soil on the lth pile in the x, y, and z directions, respectively, and are given as follows:

$$p_x^l(z) = a\int_0^{2\pi}\{[\sigma_{rr}(a,\theta_l,z) + p_f(a,\theta_l,z)]\cos\theta_l - \sigma_{r\theta}(a,\theta_l,z)\sin\theta_l\}\mathrm{d}\theta_l \tag{6.2.24}$$

$$p_y^l(z) = a\int_0^{2\pi}\{[\sigma_{rr}(a,\theta_l,z) + p_f(a,\theta_l,z)]\sin\theta_l + \sigma_{r\theta}(a,\theta_l,z)\cos\theta_l\}\mathrm{d}\theta_l \tag{6.2.25}$$

$$p_z^l(z) = a\int_0^{2\pi}\sigma_{rz}(a,\theta_l,z)\mathrm{d}\theta_l \tag{6.2.26}$$

where a is the radius of the lth pile.

Employing Eqs. (6.2.8)–(6.2.13), the wave fields at the pile-soil interfaces can be determined. Then, the stresses and pore water pressures at each interface can be obtained by introducing the wave fields into Eqs. (6.2.17)–(6.2.20). After that, taking into account the orthogonality of trigonometric functions, Eqs. (6.2.24)–(6.2.26) can be rewritten as

$$p_x^l(z) = u_1 e^{ka_1z} + u_2 e^{ka_2z} + u_3 e^{kb_1z} \tag{6.2.27}$$

$$p_y^l(z)=v_1 e^{ka_1 z}+v_2 e^{ka_2 z}+v_3 e^{kb_1 z} \tag{6.2.28}$$

$$p_z^l(z)=w_1 e^{ka_1 z}+w_2 e^{ka_2 z}+w_3 e^{kb_1 z} \tag{6.2.29}$$

where u_1, u_2, u_3, v_1, v_2, and v_3 as well as w_1, w_2, and w_3 are given in Appendix G.

The total solution of Eq. (6.2.21) is given by Eq. (6.2.30), and is the sum of the homogeneous and particular solution:

$$\begin{aligned} U_l(z)=&e^{\gamma z}(C_l^1\cos\gamma z+C_l^2\sin\gamma z)+e^{-\gamma z}(C_l^3\cos\gamma z+C_l^4\sin\gamma z)\\ &+\varepsilon_p\left[\frac{u_1 e^{ka_1 z}}{k^4a_1^4+4\gamma^4}+\frac{u_2 e^{ka_2 z}}{k^4a_2^4+4\gamma^4}+\frac{u_3 e^{kb_1 z}}{k^4b_1^4+4\gamma^4}\right]\end{aligned} \tag{6.2.30}$$

where $\gamma^4=-\dfrac{m_p\omega^2}{4E_pI_p}$, $\varepsilon_p=\dfrac{1}{E_pI_p}$. Quantities C_l^1, C_l^2, C_l^3, and C_l^4 in Eq. (6.2.30) are integration constants related to the motion of the lth pile, along the x direction, and they depend on the pile boundary conditions.

In the case of a free-head pile, at the soil surface, the bending moment is zero and the shear force is set equal to zero because the results presented herein are for the kinematic solution only. At the tip of the pile, the bending moment and the shear force are zero. Therefore, the boundary conditions can be expressed as

$$M=E_pI_pU_l''(0)=0,\ Q=E_pI_pU_l'''(0)=0 \tag{6.2.31}$$

$$M=E_pI_pU_l''(-H)=0,\ Q=E_pI_pU_l'''(-H)=0 \tag{6.2.32}$$

Let $\boldsymbol{X}$ be the matrix $[C_l^1\ C_l^2\ C_l^3\ C_l^4]^{\mathrm{T}}$ in which the superscript T indicates the transpose of the matrix. Then we can get the following relation from Eqs. (6.2.30) to (6.2.32):

$$\boldsymbol{R}_1\boldsymbol{X}=\boldsymbol{T}_1 \tag{6.2.33}$$

where

$$\boldsymbol{R}_1=\begin{bmatrix}0 & 2\gamma^2 & 0 & -2\gamma^2\\ -2\gamma^3 & 2\gamma^3 & 2\gamma^3 & 2\gamma^3\\ -2\gamma^2L_2 & 2\gamma^2L_1 & 2\gamma^2L_4 & -2\gamma^2L_3\\ -2\gamma^3(L_1+L_2) & 2\gamma^3(L_1-L_2) & 2\gamma^3(L_3-L_4) & 2\gamma^3(L_3+L_4)\end{bmatrix} \tag{6.2.34}$$

$$\boldsymbol{T}_1 = \begin{bmatrix} -\varepsilon_p\left(\dfrac{k^2a_1^2u_1}{k^4a_1^4+4\gamma^4}+\dfrac{k^2a_2^2u_2}{k^4a_2^4+4\gamma^4}+\dfrac{k^2b_1^2u_3}{k^4b_1^4+4\gamma^4}\right) \\ -\varepsilon_p\left(\dfrac{k^3a_1^3u_1}{k^4a_1^4+4\gamma^4}+\dfrac{k^3a_2^3u_2}{k^4a_2^4+4\gamma^4}+\dfrac{k^3b_1^3u_3}{k^4b_1^4+4\gamma^4}\right) \\ -\varepsilon_p\left(\dfrac{k^2a_1^2u_1\mathrm{e}^{-ka_1H}}{k^4a_1^4+4\gamma^4}+\dfrac{k^2a_2^2u_2\mathrm{e}^{-ka_2H}}{k^4a_2^4+4\gamma^4}+\dfrac{k^2b_1^2u_3\mathrm{e}^{-kb_1H}}{k^4b_1^4+4\gamma^4}\right) \\ -\varepsilon_p\left(\dfrac{k^3a_1^3u_1\mathrm{e}^{-ka_1H}}{k^4a_1^4+4\gamma^4}+\dfrac{k^3a_2^3u_2\mathrm{e}^{-ka_2H}}{k^4a_2^4+4\gamma^4}+\dfrac{k^3b_1^3u_3\mathrm{e}^{-kb_1H}}{k^4b_1^4+4\gamma^4}\right) \end{bmatrix} \tag{6.2.35}$$

where $L_1=\mathrm{e}^{-\gamma H}\cos\gamma H$, $L_2=-\mathrm{e}^{-\gamma H}\sin\gamma H$, $L_3=\mathrm{e}^{\gamma H}\cos\gamma H$, and $L_4=-\mathrm{e}^{\gamma H}\sin\gamma H$. Accordingly, constants C_l^1, C_l^2, C_l^3, and C_l^4 are determined by solving Eq. (6.2.34).

Following similar procedures, the total solution of Eq. (6.2.22) can be written as

$$V_l(z)=\mathrm{e}^{\gamma z}(D_l^1\cos\gamma z+D_l^2\sin\gamma z)+\mathrm{e}^{-\gamma z}(D_l^3\cos\gamma z+D_l^4\sin\gamma z)$$
$$+\varepsilon_p\left[\frac{v_1\mathrm{e}^{ka_1z}}{k^4a_1^4+4\gamma^4}+\frac{v_2\mathrm{e}^{ka_2z}}{k^4a_2^4+4\gamma^4}+\frac{v_3\mathrm{e}^{kb_1z}}{k^4b_1^4+4\gamma^4}\right] \tag{6.2.36}$$

where D_l^1, D_l^2, D_l^3, and D_l^4 are integration constants related to the motion of the lth pile, along the y direction, to be determined from the boundary conditions.

Suppose that $\boldsymbol{Y}=[D_l^1\ D_l^2\ D_l^3\ D_l^4]^{\mathrm{T}}$; then one has

$$\boldsymbol{R}_2\boldsymbol{Y}=\boldsymbol{T}_2 \tag{6.2.37}$$

where, $\boldsymbol{R}_2=\boldsymbol{R}_1$, and $\boldsymbol{T}_2$ = similar expression as $\boldsymbol{T}_1$ with u_1, u_2, and u_3 replaced by v_1, v_2, and v_3, respectively.

The solution of Eq. (6.2.23) is found to be

$$W_l(z)=F_l^1\mathrm{e}^{\gamma z}+F_l^2\mathrm{e}^{-\gamma z}-\varepsilon_z\left[\frac{w_1\mathrm{e}^{ka_1z}}{k^2a_1^2-\gamma_z^2}+\frac{w_2\mathrm{e}^{ka_2z}}{k^2a_2^2-\gamma_z^2}+\frac{w_3\mathrm{e}^{kb_1z}}{k^2b_1^2-\gamma_z^2}\right] \tag{6.2.38}$$

where $\gamma_z^2=-\dfrac{m_p\omega^2}{E_pA_p},\varepsilon_z=\dfrac{1}{E_pA_p}$. Constants F_l^1 and F_l^2 in Eq. (6.2.38) are determined from the boundary conditions.

Since we were interested in the kinematic response only, the vertical load at the pile head was set equal to zero. At the tip of the pile it was also assumed that the vertical load is zero. This assumption was evident when the pile was in tension. On the other hand, when the pile was in compression the load at the tip of the pile is usually very small, since the entire vertical load that floating piles carry is transmitted to the poroelastic soil through skin friction. Therefore, we can get the following relations from the previously established assumption:

$$E_pA_pW_l'(0)=0 \tag{6.2.39}$$
$$E_pA_pW_l'(-H)=0 \tag{6.2.40}$$

We define $\boldsymbol{Z}=[F_l^1\ F_l^2]^{\mathrm{T}}$; then the equations resulting from the expressions (6.2.38) to (6.2.40) can be cast in a matrix form as

$$\boldsymbol{R}_3\boldsymbol{Z}=\boldsymbol{T}_3 \tag{6.2.41}$$

where

$$\boldsymbol{R}_3=\begin{bmatrix}\gamma_z & -\gamma_z\\ \gamma_z \mathrm{e}^{-\gamma_z H} & -\gamma_z \mathrm{e}^{\gamma_z H}\end{bmatrix} \tag{6.2.42}$$

$$\boldsymbol{T}_3=\begin{bmatrix}\varepsilon_z\left(\dfrac{ka_1w_1}{k^2a_1^2-\gamma_z^2}+\dfrac{ka_2w_2}{k^2a_2^2-\gamma_z^2}+\dfrac{kb_1w_3}{k^2b_1^2-\gamma_z^2}\right)\\ \varepsilon_z\left(\dfrac{ka_1w_1\mathrm{e}^{-ka_1H}}{k^2a_1^2-\gamma_z^2}+\dfrac{ka_2w_2\mathrm{e}^{-ka_2H}}{k^2a_2^2-\gamma_z^2}+\dfrac{kb_1w_3\mathrm{e}^{-kb_1H}}{k^2b_1^2-\gamma_z^2}\right)\end{bmatrix} \tag{6.2.43}$$

The appropriate boundary conditions to be satisfied at the pile-soil interfaces are continuity of the solid skeleton displacements, i.e.,

$$u_r\,(a,\theta_l,z)=U_l\,(z)\cos\theta_l+V_l\,(z)\sin\theta_l \tag{6.2.44}$$

$$u_\theta\,(a,\theta_l,z)=-U_l\,(z)\sin\theta_l+V_l\,(z)\cos\theta_l \tag{6.2.45}$$

$$u_z\,(a,\theta_l,z)=W_l\,(z) \tag{6.2.46}$$

where $0\le\theta_l\le 2\pi$; $-H\le z\le 0$; $l=1,2,\ldots,N$.

Utilizing Eqs. (6.2.44)–(6.2.46) and taking into account the linear independence of trigonometric functions, then six infinite linear systems of algebraic equations for the scattering coefficients A_n^j through B_n^j can be obtained:

$$\frac{\varepsilon_m}{2}\sum_{n=0}^{\infty}\left[(1-\delta_{l1})\sum_{j=1}^{l-1}(-1)^m A_n^j K_m^n(kd_{jl})+(1-\delta_{lN})\sum_{j=l+1}^{N}(-1)^n A_n^j K_m^n(kd_{jl})\right]\cdot\mathfrak{R}_q^{11}$$
$$+A_m^l\cdot\mathfrak{R}_q^{12}=-\varepsilon_m(-\mathrm{i})^m\cos\frac{m\pi}{2}\cdot\mathfrak{R}_q^{11}+\mathfrak{R}_q^{13} \tag{6.2.47}$$

$$\frac{\varepsilon_m}{2}\sum_{n=0}^{\infty}\left[(1-\delta_{l1})\sum_{j=1}^{l-1}(-1)^m B_n^j L_m^n(kd_{jl})+(1-\delta_{lN})\sum_{j=l+1}^{N}(-1)^n B_n^j L_m^n(kd_{jl})\right]\cdot\mathfrak{R}_q^{21}$$
$$+B_m^l\cdot\mathfrak{R}_q^{22}=-\varepsilon_m(-\mathrm{i})^m\sin\frac{m\pi}{2}\cdot\mathfrak{R}_q^{21}+\mathfrak{R}_q^{23} \tag{6.2.48}$$

where δ_{lN} is the Kronecker delta, $\delta_{lN}=1$ if $l=N$, $\delta_{lN}=0$ if $l\neq N$; $q=1,2,3$; $m=0,1,2,\ldots,\infty$; and

$\mathfrak{R}_q^{11}$ through $\mathfrak{R}_q^{23}$ are the expressions related to the Bessel function and the Hankel function.

These six infinite linear systems of algebraic equations for the scattering coefficients A_n^j through B_n^j cannot be solved in an exact form. They need to be truncated with an appropriate range for the expansions and need to be solved in the complex least-squares sense. Throughout this section, $C^{m\times n}$ denotes the linear vector space of m-by-n complex matrices. Now, we consider the following linear system:

$$\boldsymbol{RQ}=\boldsymbol{G} \tag{6.2.49}$$

where $\boldsymbol{R}=[r_{ij}]\in C^{m\times n}$ $(m > n)$ and $\boldsymbol{G}\in C^n$ are given, and $\boldsymbol{Q}\in C^n$ is an unknown vector.

A least-squares approximate solution to the complex system Eq. (6.2.49) can be derived as

$$\hat{\boldsymbol{Q}}=(\bar{\boldsymbol{R}}^{\mathrm{T}}\boldsymbol{R})^{-1}\,\boldsymbol{R}^{\mathrm{T}}\boldsymbol{G} \tag{6.2.50}$$

where $\bar{\boldsymbol{R}}^{\mathrm{T}}$ stands for the conjugate transpose of the matrix $\boldsymbol{R}$, and $(\cdot)^{-1}$ stands for the inverse of the matrix.

Once the systems of Eqs. (6.2.47), (6.2.48) are truncated with an appropriate range for the expansions and solved with the aid of the expression Eq. (6.2.50), the scattering coefficients can be subsequently obtained. Then, the corresponding displacements, stresses and pore water pressures in each point of the poroelastic soil can be calculated.

6.2.3 Numerical Results and Conclusions

The piles were assumed to be identical and equally spaced with a radius a, pile length H, mass density ρ_{p}, and pile spacing sp, where sp denotes the distance between the centers of adjacent piles. In addition, the origin of the rectangular coordinate system was placed at the center of the barrier as depicted in Fig. 6.2.2. The normalized displacement amplitude $|u_z/u_0|$ was defined as the ratio of the solid displacement amplitude of soil along the z direction in the presence of the barrier over that displacement amplitude in the absence of the barrier.

It is convenient at this stage to introduce dimensionless variables as in Section 6.1.2.

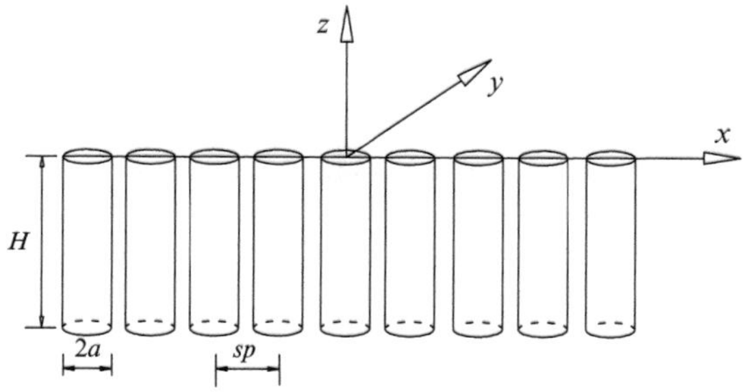

Fig. 6.2.2 A row of piles and a rectangular coordinate system

Comparison with Published Results

It is noted that if poroelastic parameters ρ_f^*, K_f^*, and b^* approach zero, then the poroelastic medium is degenerated into an elastic medium. In this section, results obtained based on the present methodology are compared with those of Tsai et al.[45] The material properties of the elastic soil were chosen as: shear modulus μ_s = 132 MPa, mass density ρ_s = 1750 kg/m^3, Poisson's ratio v_s = 0.25, and hysteretic damping β_s = 5%. The properties of the piles are shear modulus $\mu_p = 34.29\mu_s$, mass density $\rho_p = 1.37\rho_s$, and Poisson's ratio $v_p = v_s$. The wavelength of incident Rayleigh waves λ_R for the soil was assumed to be 5 m and the geometry parameters of a row of eight piles in dimensionless terms reads a/λ_R = 0.1, H/λ_R = 1.0, and sp/λ_R = 0.3. Fig. 6.2.3 shows the variation of the normalized vertical displacement amplitude $|u_z/u_0|$ with y/λ_R behind a pile barrier. Slight differences can be seen between the two results. This discrepancy could be due to the fact that the present method only considers the incidence of Rayleigh waves, while Tsai et al.[45] employed the boundary element method to deal with all kinds of waves.

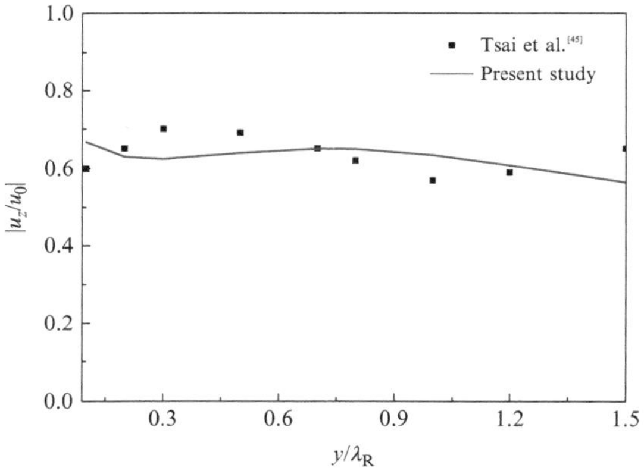

Fig. 6.2.3 Comparison of amplitude reduction of Rayleigh waves by a row of piles

Parametric Study

Some numerical examples are presented to investigate the influence of certain parameters on the screening of Rayleigh waves achieved by piles in a row. The poroelastic soil and pile properties are summarized in Table 6.2.1. Unless otherwise stated, the length of the piles H, pile spacing sp, soil depth z, and distance y away from the pile barrier are normalized with respect to the wavelength of incident Rayleigh waves in poroelastic soil with b^* = 100 when ω^* = 1.2.

Table 6.2.1 Input parameters in the calculation

Parameter	Value
λ^*	1.004
K_f^*	76.63
f	0.27
ρ_f^*	0.454
ρ_v^*	1.134
E_v^*	1000

Fig. 6.2.4 presents the variation of $|u_z/u_0|$ with y/λ_R (λ_R is the wavelength of incident Rayleigh waves when $\omega^* = 1.2$, $b^* = 100$) in poroelastic soil for five different dimensionless frequencies $\omega^* =$ 0.6, 0.9, 1.2, 1.5, and 1.8, respectively. For each dimensionless frequency, there is an optimum value of y/λ_R that gives the lowest value of $|u_z/u_0|$. This is called the optimum screening location. It ranges from 2.8 for $\omega^* = 1.2$ to 5.8 for $\omega^* = 1.8$. On the other hand, when the dimensionless frequency is low ($\omega^* = 0.6$), the pile barrier has only a very small effect on the isolation of Rayleigh waves. As the dimensionless frequency increases, the optimum screened position is moved backward, while the minimum value of $|u_z/u_0|$ decreases; that is to say, the isolation effectiveness of the pile barrier increases. For the five dimensionless frequencies considered, the maximum variation in the normalized displacement amplitude $|u_z/u_0|$ was 82%. All these indicate that there is a certain frequency range for a row of piles to isolate the Rayleigh wave, and the larger the frequency of an incident wave, the better the isolation effectiveness of a pile barrier. This result is physically reasonable, as the interaction between incident and scattered waves will strengthen when the frequency of incident Rayleigh waves increases.

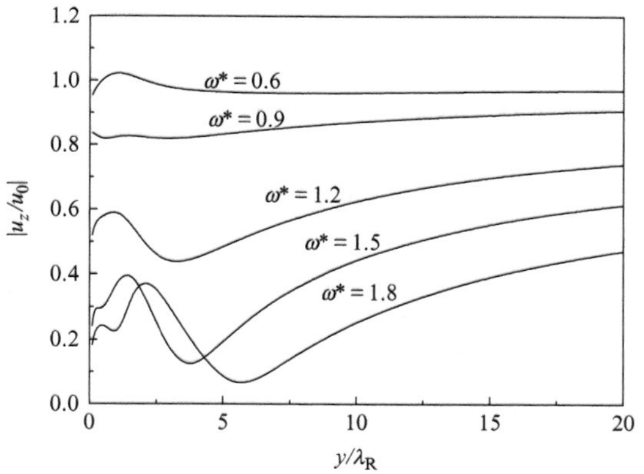

Fig. 6.2.4 Comparsion of $|u_z/u_0|$ under different dimensionless frequencies ($N = 9$, $sp/\lambda_R = 0.32$, $H/\lambda_R = 2$, $x/\lambda_R = 0$, $z/\lambda_R = 0$, $b^* = 100$)

As expected, an increase in the length of the piles resulted in a better performance of the pile barrier. Fig. 6.2.5 shows the plotting of $|u_z/u_0|$ versus y/λ_R under various pile lengths. When H/λ_R varied from 1.0 to 3.0, the optimum isolation efficiency of the pile barrier was ranged from 36% to 74%. To achieve a normalized displacement amplitude of approximately 0.44 or less, a normalized length H/λ_R of roughly 2.0 was needed. Increasing length, generally speaking, means blocking more of the incident Rayleigh waves and appears to contribute to a better screening. However, due to the nature of Rayleigh waves, which are confined near the surface of a poroelastic half-space, this effect will be less important for very long piles.

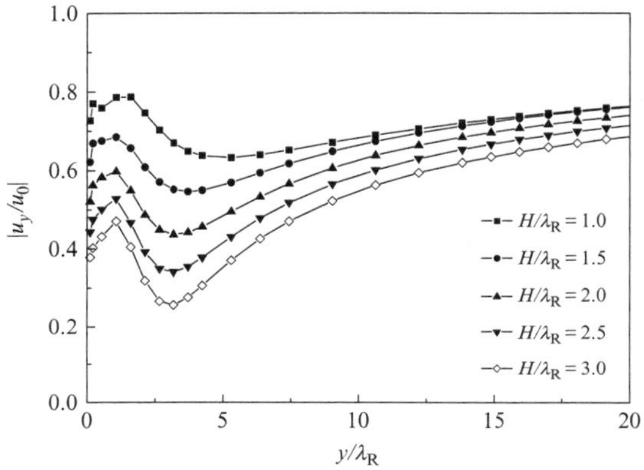

Fig. 6.2.5 $|u_z/u_0|$ versus y/λ_R under different pile lengths ($\omega^* = 1.2$, $N = 9$, $sp/\lambda_R = 0.32$, $b^* = 100$, $x/\lambda_R = 0$, $z/\lambda_R = 0$)

Fig. 6.2.5 also show an unusual trend for the case $y/\lambda_R < 2.5$, which indicates the presence of strong scattered waves of similar wavelength traveling in the direction of the incident Rayleigh waves and thereby causing constructive or destructive interference, due to their phase difference. This notion also helps one to explain the rise in all the curves after reaching a minimum.

The influence of the number of piles on the isolation of Rayleigh waves is shown in Fig. 6.2.6. An increase in the number of piles offers greater resistance (as a whole obstacle) to the otherwise free-field motion of the incoming Rayleigh wave and results in a better screening, but the optimum isolation location moves backwards at a certain distance. Another point to note is that the optimum isolation performance of the pile barrier increases by only 4% when the number of piles is varied from 6 to 12. This implies that increasing the number of piles can give a greater scope to the optimum screened zone, but cannot remarkably reduce the displacement amplitude.

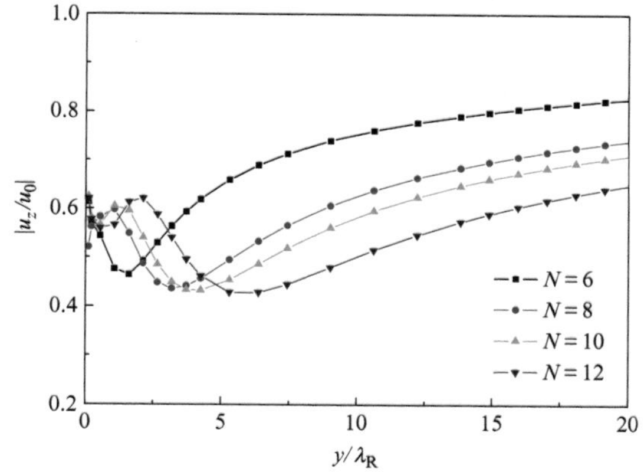

Fig. 6.2.6 $|u_z/u_0|$ versus y/λ_R under different numbers of piles ($\omega^* = 1.2$, $sp/\lambda_R = 0.32$, $H/\lambda_R = 2$, $b^* = 100$, $x/\lambda_R = 0$, $z/\lambda_R = 0$)

Fig. 6.2.7 depicts the variation of $|u_z/u_0|$ with y/λ_R at different soil depths behind a pile barrier. The normalized soil depth z/λ_R of 0 (at the ground surface) may result in a maximum amplitude reduction of 0.56. The value of $|u_z/u_0|$ increases with a decrease in normalized depth for the same distance away from the barrier when $y/\lambda_R<8$. This implies that the displacement amplitude reduction in deeper soil works against the purpose of a pile barrier. The trend of the curve for $z/\lambda_R=-2.0$ is very close to that for $z/\lambda_R=-1.5$; that is, increased soil depth contributes almost nothing toward the screening efficiency, which suggests that the propagation depth of Rayleigh waves is about 1.5 times the wavelength.

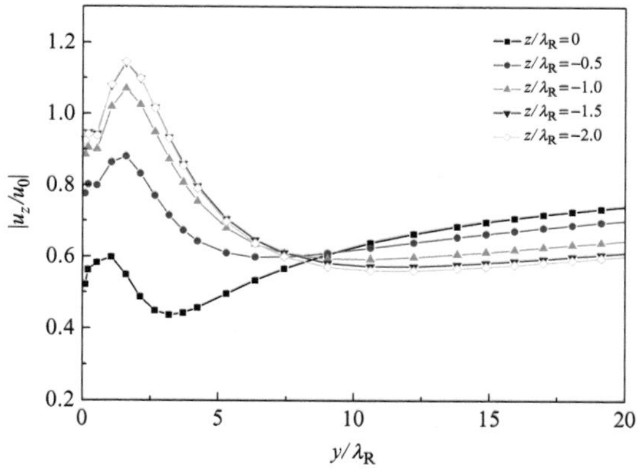

Fig. 6.2.7 $|u_z/u_0|$ versus y/λ_R under different soil depths ($\omega^* = 1.2$, $sp/\lambda_R = 0.32$, $H/\lambda_R = 2$, $b^* = 100$, $x/\lambda_R = 0$, $N = 9$)

Fig. 6.2.8 illustrates the variation of the normalized pore water pressure in the presence of a pile barrier (normalized with respect to the maximum pore water pressure at $y/\lambda_R=3.2$ in the absence of a pile barrier). From the observation of the pore water pressure curves, pressure evidently declines with soil depth and reduces to nearly zero at z/λ_R=−1.5. Fig. 6.2.8 also indicates that a decrease in pile spacing can significantly reduce the pore water pressure in poroelastic soil after excitation by the passage of Rayleigh waves.

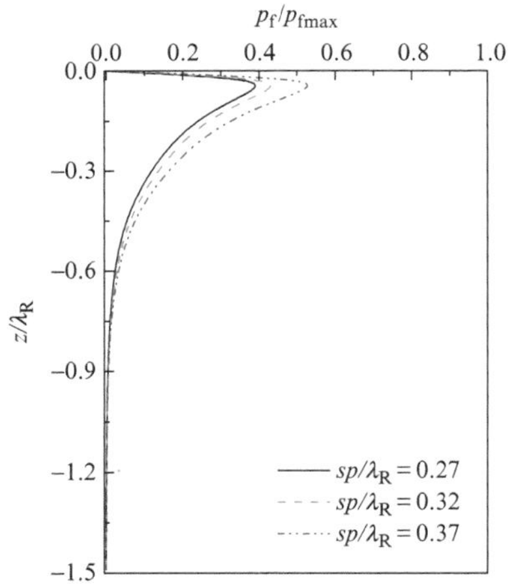

Fig. 6.2.8 Variation of normalized pore water pressure with z/λ_R after installation of a pile barrier ($\omega^* = 1.2$, $H/\lambda_R = 2$, $x/\lambda_R = 0$, $N = 9$, $b^* = 100$, $y/\lambda_R = 3.2$)

Fig. 6.2.9 shows the contour diagrams of $|u_z/u_0|$ for poroelastic soil. It suggests that the relatively small areas immediately close to the rows of piles are inclined to produce displacement fluctuations. In the center of the screened zone, the optimum isolation location does not appear very close to the pile barrier but at a certain distance away from it.

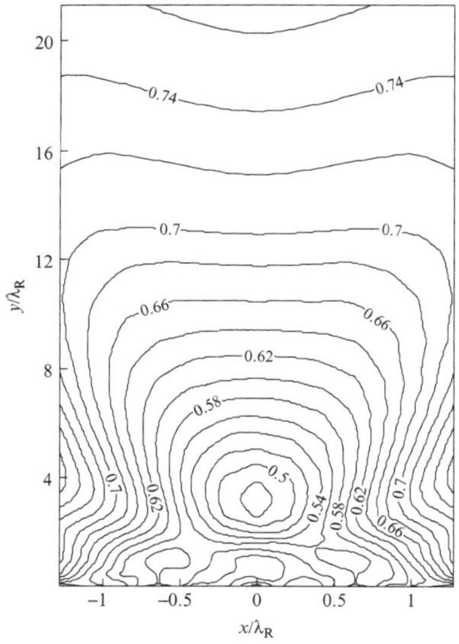

Fig. 6.2.9 Contours of $|u_z/u_0|$ for poroelastic soil ($\omega^* = 1.2$, $sp/\lambda_R = 0.32$, $H/\lambda_R = 2$, $N = 9$, $b^* = 1000$)

6.2.4 Conclusions

The isolation of Rayleigh waves by a row of piles in a poroelastic medium was investigated. The influence of certain soil and pile characteristics, such as the permeability of poroelastic soil, spacing between the piles and length of the piles on the isolation performance of a pile barrier, was studied. Based on the results presented herein, the following conclusions can be drawn for the isolation of Rayleigh waves in poroelastic soil:

(1) In poroelastic soil, an increased length of the piles appears to contribute to a better isolation. However, this effect will be less important for very long piles due to the nature of Rayleigh waves, which can only be propagated within a certain depth of a soil stratum.

(2) The pore water pressure declines evidently with soil depth, and the presence of a pile barrier in poroelastic soil can significantly reduce the pore water pressure when excited by the passage of Rayleigh waves.

Chapter 7
Biot's Theory in the Finite Element Method

Since 1975 a large amount of research work, both at Swansea and elsewhere around the world, has been directed toward laying the foundations for numerical predictions based largely on concepts concerning the dynamic interactions between the two phases of soil—i.e., the soil skeleton phase and the pore water phase—that were proposed by Biot in the 1940s. Notably, the researchers at Swansea under the leadership of Zienkiewicz et al.[3] made significant contributions in developing finite element method (FEM) codes (SWANDYNE), by adopting different formulations of Biot's porodynamic theory and incorporating advanced elasto-plastic models, to capture the phenomena of the weakening and liquefaction in soil when subjected to repeated loading, such as in earthquakes. To fully test the possibilities offered by the FEM formulation, the computational predictions on the pore pressure and the permanent displacement generations under the artificial earthquake wave inputs were compared with results from centrifuge experiments, e.g., the VELACS project. In the following 20 years, the center of development of computational tools tackling earthquake-induced engineering problems, focusing on structure collapse and ground liquefactions, has shifted from Europe to the United States, landmarked by the establishment of the Pacific Earthquake Engineering Research Center (PEER) in California in 1997. PEER has developed an Open System for Earthquake Engineering Simulation, OpenSees (http://opensees.berkeley.edu) for short, as a software platform to simulate the performance of structural and geotechnical systems subjected to earthquakes. OpenSees is an open source computational platform that is under constant development, contributed to by universities and individuals both in the United States and around the world. It has incorporated many advanced capabilities for modeling and analyzing the nonlinear response of systems using a wide range of material models, elements, and solution algorithms. For example, a pressure independent/dependent multiyield material model, considering both the soil skeleton and the pore water phases, was contributed by the University of California at San Diego for nonlinear soil responses under general 2D/3D cyclic loading conditions.

Unlike the long-lasting research effort on the soil dynamic responses induced by earthquake loads, the contemporary research on soil behaviors under repetitive traffic loads has received limited attention, although both earthquake and traffic loads are cyclic loads in nature. It was not until the last two decades that the dynamic soil responses under traffic loads raised much attention (most of the investigations employ an analytical or semianalytical scheme), when a worldwide deployment of high-speed trains was seen. Compared to the earthquake load, the traffic load has its own characteristics: (1) it is moving; (2) it is applied at a higher frequency and lasts during the entire serving life of the engineering structure; (3) its amplitude is generally low so that the soil under the embankment will behave linearly. Thus, instead of implementing advanced nonlinear material models, the inertia effects both in the soil skeleton and the pore water phases are of considerable importance and should be fully taken into account, whereas it is a common practice in earthquake computations to neglect the inertia term of the pore water under the observation that earthquake loading is low frequency in nature. Moreover, proper artificial boundary conditions should be proposed to accommodate the moving load and to damp the associating outgoing wave fields.

Due to these considerations, this chapter has been arranged as follows: Section 7.1 demonstrates the formulation and the verification of the proposed saturated soil elements based on the complete Biot's formulations (i.e., the ***u-w*** formulation); Section 7.2 proposes a novel transmitting boundary condition for the ***u-w*** formulation, whose efficiency was analytically derived and computationally verified; based on the results of the two sections, Section 7.3 draws some conclusions along with suggested topics for future research.

7.1 Formulation of the Saturated Soil Elements

The original porodynamic theory given by Biot takes the soil skeleton displacement $\boldsymbol{u}$ and the pore-fluid average displacement relative to the soil skeleton $\boldsymbol{w}$ as primary unknowns. Based on his theory, the following research workers adopted different formulations and simplifications. To study the two-dimensional (2D) wave propagation problem caused by seismic or blast load, Prevost[46] formulated a finite element procedure using the soil displacement $\boldsymbol{u}$ and the pore-fluid relative velocity $\dot{\boldsymbol{w}}$ as unknowns. The time integration of the resulting semidiscrete finite element equations was carried out by using a staggered implicit-explicit algorithm. By neglecting the pore-fluid relative acceleration term $\ddot{\boldsymbol{w}}$, Zienkiewicz et al.[3] formulated a saturated soil element based on soil displacement $\boldsymbol{u}$ and excess pore water pressure p as primary variables. It was successfully applied in liquefaction predictions of sandy soil under seismic load. By incorporating an advanced multiyield material model for the soil skeleton, Elgamal et al.[47–48] adopted the same ***u-p*** formulation to investigate ground liquefactions. However, an analytical research by Zienkiewicz[49] indicated that obvious errors would occur in dynamic responses of soils involving high permeability or high-frequency excitations when the pore-fluid relative acceleration term $\ddot{\boldsymbol{w}}$ was neglected. Using the Galerkin method, Gajo et al.[50] derived three finite element equations for the ***u-w***, ***u-p***, and ***u-p-U*** formulations, respectively, where $\boldsymbol{U}$ denotes the absolute displacement vector of the pore fluid. The ***u-p-U*** formulation suits the limit situation when both the pore fluid and the soil grains are incompressible. Based on the variation principle, Zhang[42] derived the finite element equation using the ***u-U*** formulation that was adopted to study the vibration problem of a rigid foundation resting on a saturated ground. Most recently, Jeremic et al.[43] developed a three-dimensional (3D) finite element procedure using the ***u-p-U*** formulation to study the liquefactions of sandy soil. It was noted that the simplified ***u-p*** formulation had the fewest nodal variables (3+1) compared with the number of nodal unknowns associated with the complete ***u-w*** and ***u-U*** formulations (3+3), whereas the ***u-p-U*** formulation had the most nodal variables (3+1+3).

For the dynamic responses of a saturated ground caused by traffic loads (e.g., loads of vehicles and trains), there has been a limited research effort employing a finite element scheme. Ketti et al.[44–45] studied the dynamic responses of 2D saturated ground under a moving train axle load using a ***u-p-w*** formulation. Yang and Jin[46] studied the 3D dynamic responses of poroelastic soil subjected to a high-speed load using the moving element method established for the ***u-p*** formulation. By decomposing the wave fields into a Fourier series along the load-moving direction, the 3D ground responses could be studied using a 2D FEM mesh. This scheme, which is a novel combination of numerical and analytical methods and greatly reduced the computation load, is called the two-and-one-half FEM (2.5D FEM). Gao et al.[47] established a 2.5D FEM procedure for the ***u-w*** formulation to study the dynamic responses of saturated ground under a moving point load. However, the Fourier series decomposition involved in the 2.5D FEM scheme implies that the constitutive behavior should be linear and the cross-section of the ground should be invariant along the longitudinal direction (i.e., the load-moving direction). This hinders its application in soils associated with a nonlinear behavior or nonhomogeneity.

Since the simplified ***u-p*** formulation would cause errors when the load velocity, the load oscillating frequency or the soil permeability is high, this section derives the finite element equations for the saturated soil associated with a compressible pore fluid based on the complete ***u-w*** formulation. The ***u-U*** formulation was not chosen due to the consideration that the natural boundary conditions corresponding to the nodal variables ***u*** and ***U*** are the effective stress and the pore-fluid stress (equal to the multiplication of the pore-fluid pressure and the porosity), respectively. The implementation of these natural boundary conditions was not as straightforward as the implementation of those associated with the ***u-w*** formulation, i.e., the total stress and the pore-fluid pressure.

7.1.1 Spatial Discretization of Finite Element Equations

The complete ***u-w*** formulation governing the dynamic responses of the saturated half-space was given in Eqs. (1.1.24), (1.1.25) in Chapter 1. Denoting the FEM solution domain as Ω, the boundary conditions specified on its boundary Γ are shown as follows.

The essential boundary conditions:

$$\begin{cases} u_i = \bar{u}_i \text{ on } \Gamma = \Gamma_{\mathrm{D}}^{\mathrm{s}} \\ w_i = \bar{w}_i \text{ on } \Gamma = \Gamma_{\mathrm{D}}^{\mathrm{w}} \end{cases} \tag{7.1.1}$$

The natural boundary conditions:

$$\begin{cases} \sigma_{ij} n_j = \bar{T}_i \text{ on } \Gamma = \Gamma_{\mathrm{N}}^{\mathrm{t}} \\ p = \bar{p} \text{ on } \Gamma = \Gamma_{\mathrm{N}}^{\mathrm{p}} \end{cases} \tag{7.1.2}$$

and

$$\begin{cases} \Gamma = \Gamma_{\mathrm{D}}^{\mathrm{s}} \cup \Gamma_{\mathrm{D}}^{\mathrm{w}} \cup \Gamma_{\mathrm{N}}^{\mathrm{t}} \cup \Gamma_{\mathrm{N}}^{\mathrm{p}} \\ \Gamma_{\mathrm{D}}^{\mathrm{s}} \cap \Gamma_{\mathrm{N}}^{\mathrm{t}} = \varnothing \\ \Gamma_{\mathrm{D}}^{\mathrm{w}} \cap \Gamma_{\mathrm{N}}^{\mathrm{p}} = \varnothing \end{cases} \tag{7.1.3}$$

We introduced shape functions of the same order for the field variables ***u*** and ***w***, respectively. The spatial discretization is written as

$$\begin{cases} \boldsymbol{u}^{\mathrm{e}}(\boldsymbol{x},t)=\sum_{i=1}^{n^{\mathrm{e}}} N_i(\boldsymbol{x})\boldsymbol{I}\hat{\boldsymbol{u}}_i^{\mathrm{e}}(t) \\ \boldsymbol{w}^{\mathrm{e}}(\boldsymbol{x},t)=\sum_{i=1}^{n^{\mathrm{e}}} N_i(\boldsymbol{x})\boldsymbol{I}\hat{\boldsymbol{w}}_i^{\mathrm{e}}(t) \end{cases} \tag{7.1.4}$$

where $\boldsymbol{u}^{\mathrm{e}}(\boldsymbol{x}, t)$ and $\boldsymbol{w}^{\mathrm{e}}(\boldsymbol{x}, t)$ denote the soil-skeleton displacement and the pore-fluid relative displacement within the element, respectively; $\boldsymbol{x}=(x, y, z)$ is the location vector; t is time; $\hat{u}_i^{\mathrm{e}}(t)=\{\hat{u}_{i1}^{\mathrm{e}}\ldots\hat{u}_{id}^{\mathrm{e}}\}^{\mathrm{T}}$ and $\boldsymbol{w}_i^{\mathrm{e}}(t)=\{\hat{w}_{i1}^{\mathrm{e}}\ldots\hat{w}_{id}^{\mathrm{e}}\}^{\mathrm{T}}$ are the nodal displacement vectors associated with the soil skeleton and the pore fluid, respectively; $N_i(\boldsymbol{x})$ $(i=1, 2,\ldots, n^{\mathrm{e}})$ is the shape function; n^{e} is the number of nodes; $\boldsymbol{I}$ is an identity matrix of dimension $d\times d$; $d=2$ and 3 denote a 2D and 3D element, respectively.

It was assumed that the interpolation is such that the essential boundary conditions, Eq. (7.1.1), are satisfied on $\Gamma_{\mathrm{D}}^{\mathrm{s}}$ and $\Gamma_{\mathrm{D}}^{\mathrm{w}}$ automatically by a suitable prescription of the nodal parameters. And the natural boundary conditions Eq. (7.1.2) will be obtained by integrating the weighted equation by parts after introducing the expansion in Eq. (7.1.4) into the governing equations, Eqs. (1.1.24), (1.1.25). After this, the original governing equations in a differential form were transformed into a weak integral form, given as

$$\begin{cases} \int_{\Omega}\boldsymbol{B}^{\mathrm{T}}\boldsymbol{\sigma}''\,\mathrm{d}\Omega+\boldsymbol{K}'_{\mathrm{ss}}\hat{\boldsymbol{u}}^{\mathrm{e}}+\boldsymbol{K}_{\mathrm{sw}}\hat{\boldsymbol{w}}^{\mathrm{e}}+\boldsymbol{M}_{\mathrm{ss}}\ddot{\hat{\boldsymbol{u}}}^{\mathrm{e}}+\boldsymbol{M}_{\mathrm{sw}}\ddot{\hat{\boldsymbol{w}}}^{\mathrm{e}}=\boldsymbol{f}^{\mathrm{s}} \\ \boldsymbol{K}_{\mathrm{ws}}\hat{\boldsymbol{u}}^{\mathrm{e}}+\boldsymbol{K}_{\mathrm{ww}}\hat{\boldsymbol{w}}^{\mathrm{e}}+\boldsymbol{M}_{\mathrm{ws}}\ddot{\hat{\boldsymbol{u}}}^{\mathrm{e}}+\boldsymbol{M}_{\mathrm{ww}}\ddot{\hat{\boldsymbol{w}}}^{\mathrm{e}}+\boldsymbol{C}_{\mathrm{ww}}\dot{\hat{\boldsymbol{w}}}^{\mathrm{e}}=\boldsymbol{f}^{\mathrm{w}} \end{cases} \tag{7.1.5}$$

where $\boldsymbol{u}^{\mathrm{e}}=\{\hat{\boldsymbol{u}}_1^{\mathrm{e}}\ldots\hat{\boldsymbol{u}}_{n^{\mathrm{e}}}^{\mathrm{e}}\}^{\mathrm{T}}$ and $\hat{\boldsymbol{w}}^{\mathrm{e}}=\{\hat{\boldsymbol{w}}_1^{\mathrm{e}}\ldots\hat{\boldsymbol{w}}_{n^{\mathrm{e}}}^{\mathrm{e}}\}^{\mathrm{T}}$ are vectors assembling all the nodal displacements of the element corresponding to the soil-skeleton and the pore-fluid phases, respectively; $\int_{\Omega}\boldsymbol{B}^{\mathrm{T}}\boldsymbol{\sigma}''\,\mathrm{d}\Omega$ denotes the reaction force provided by the soil skeleton. Specifically, for a linear elastic material we have $\int_{\Omega}\boldsymbol{B}^{\mathrm{T}}\boldsymbol{D}\boldsymbol{B}\mathrm{d}\Omega\hat{\boldsymbol{u}}^{\mathrm{e}}=\boldsymbol{K}_{\mathrm{ss}}\hat{\boldsymbol{u}}^{\mathrm{e}}$ by substituting $\boldsymbol{\sigma}''=\boldsymbol{D}\boldsymbol{B}\hat{\boldsymbol{u}}^{\mathrm{e}}$. For an elasto-plastic material it takes the following incremental form: $\mathrm{d}\boldsymbol{\sigma}''=(\boldsymbol{D}-\boldsymbol{D}_{\mathrm{p}})\,\mathrm{d}\boldsymbol{\varepsilon}=(\boldsymbol{D}-\boldsymbol{D}_{\mathrm{p}})\boldsymbol{B}\mathrm{d}\hat{\boldsymbol{u}}^{\mathrm{e}}$; $\boldsymbol{K}'_{\mathrm{ss}}=\int_{\Omega}\boldsymbol{B}^{\mathrm{T}}\boldsymbol{m}\alpha^2M\boldsymbol{m}^{\mathrm{T}}\boldsymbol{B}\mathrm{d}\Omega$ is an additional stiffness matrix of the soil skeleton; $\boldsymbol{K}_{\mathrm{sw}}=\boldsymbol{K}_{\mathrm{ws}}=\int_{\Omega}\boldsymbol{B}^{\mathrm{T}}\boldsymbol{m}\alpha M\boldsymbol{m}^{\mathrm{T}}\boldsymbol{B}\mathrm{d}\Omega$ is the coupled stiffness matrix between the soil skeleton and the pore fluid; $\boldsymbol{K}_{\mathrm{ww}}=\int_{\Omega}\boldsymbol{B}^{\mathrm{T}}\boldsymbol{m}M\boldsymbol{m}^{\mathrm{T}}\boldsymbol{B}\mathrm{d}\Omega$ is the stiffness matrix of the pore fluid; $\boldsymbol{M}_{\mathrm{ss}}=\int_{\Omega}\boldsymbol{N}^{\mathrm{T}}\rho\boldsymbol{N}\mathrm{d}\Omega$ is the mass matrix of the soil skeleton; $\boldsymbol{M}_{\mathrm{sw}}=\boldsymbol{M}_{\mathrm{ws}}=\int_{\Omega}\boldsymbol{N}^{\mathrm{T}}\rho_{\mathrm{f}}\boldsymbol{N}\mathrm{d}\Omega$ is the coupling mass matrix between the soil skeleton and the pore fluid; $\boldsymbol{M}_{\mathrm{ww}}=\int_{\Omega}\boldsymbol{N}^{\mathrm{T}}m\boldsymbol{N}\mathrm{d}\Omega$ is the mass matrix of the pore fluid; $\boldsymbol{C}_{\mathrm{ww}}=\int_{\Omega}\boldsymbol{N}^{\mathrm{T}}b\boldsymbol{N}\mathrm{d}\Omega$ denotes the damping matrix contributed by the relative motion between the two phases; $\boldsymbol{f}^{\mathrm{s}}=\int_{\Omega}\boldsymbol{N}^{\mathrm{T}}\rho\boldsymbol{b}\mathrm{d}\Omega+\int_{\Gamma_{\mathrm{N}}^{\mathrm{t}}}\boldsymbol{N}^{\mathrm{T}}\bar{\boldsymbol{T}}\mathrm{d}\Gamma$ and $\boldsymbol{f}^{\mathrm{w}}=\int_{\Omega}\boldsymbol{N}^{\mathrm{T}}\rho_{\mathrm{f}}\,\boldsymbol{b}\mathrm{d}\Omega-\int_{\Gamma_{\mathrm{N}}^{\mathrm{p}}}\boldsymbol{N}^{\mathrm{T}}\bar{p}\boldsymbol{n}\mathrm{d}\Gamma$ are the load vectors associated with the soil skeleton and the pore fluid, respectively; $\boldsymbol{\sigma}''=\{\sigma_x''\ \sigma_y''\ \sigma_z''\ \tau_{xy}\ \tau_{yz}\ \tau_{zx}\}^{\mathrm{T}}$ is the effective stress vector of the soil skeleton; $\boldsymbol{\varepsilon}=\{\varepsilon_x\ \varepsilon_y\ \varepsilon_z\ \gamma_{xy}\ \gamma_{yz}\ \gamma_{zx}\}^{\mathrm{T}}$ is the strain vector; $\boldsymbol{B}=\boldsymbol{LN}$ and $\boldsymbol{N}=\{N_1\boldsymbol{I}\ldots N_{n^e}\boldsymbol{I}\}$ is the shape function; $\boldsymbol{L}$ is the manipulator matrix that relates the strain and displacement vectors; $\boldsymbol{D}$ is the elastic constitutive matrix relating the stress and the strain vectors; $\boldsymbol{D}^{\mathrm{p}}$ is the elasto-plastic constitutive matrix that is dependent on the stress status and history; $\boldsymbol{m}=\{1\ 1\ 1\ 0\ 0\ 0\}^{\mathrm{T}}$.

7.1.2 Temporal Discretization of Finite Element Equations

The semidiscrete finite element equations obtained here are ordinary differential equations in time. To complete the numerical solution, they were integrated using the Newmark time marching scheme. Suppose the nodal kinematic values (i.e., displacement, velocity, and acceleration) are obtained at time $n\Delta t$; then the nodal velocities and accelerations at time $(n+1)\Delta t$ are given:
For the soil skeleton:

$$\begin{cases} \dot{\hat{\boldsymbol{u}}}^{\mathrm{e}}_{n+1} = \dfrac{\gamma}{\beta\Delta t}\left(\hat{\boldsymbol{u}}^{\mathrm{e}}_{n+1} - \hat{\boldsymbol{u}}^{\mathrm{e}}_{n}\right) + \left(1-\dfrac{\gamma}{\beta}\right)\dot{\hat{\boldsymbol{u}}}^{\mathrm{e}}_{n} + \left(1-\dfrac{\gamma}{2\beta}\right)\Delta t\ddot{\hat{\boldsymbol{u}}}^{\mathrm{e}}_{n} \\ \ddot{\hat{\boldsymbol{u}}}^{\mathrm{e}}_{n+1} = \dfrac{1}{\beta\Delta t^2}\left(\hat{\boldsymbol{u}}^{\mathrm{e}}_{n+1} - \hat{\boldsymbol{u}}^{\mathrm{e}}_{n}\right) - \dfrac{1}{\beta\Delta t}\dot{\hat{\boldsymbol{u}}}^{\mathrm{e}}_{n} - \left(\dfrac{1}{2\beta}-1\right)\ddot{\hat{\boldsymbol{u}}}^{\mathrm{e}}_{n} \end{cases} \tag{7.1.6}$$

For the pore fluid:

$$\begin{cases} \dot{\hat{\boldsymbol{w}}}^{\mathrm{e}}_{n+1} = \dfrac{\gamma}{\beta\Delta t}\left(\hat{\boldsymbol{w}}^{\mathrm{e}}_{n+1} - \hat{\boldsymbol{w}}^{\mathrm{e}}_{n}\right) + \left(1-\dfrac{\gamma}{\beta}\right)\dot{\hat{\boldsymbol{w}}}^{\mathrm{e}}_{n} + \left(1-\dfrac{\gamma}{2\beta}\right)\Delta t\ddot{\hat{\boldsymbol{w}}}^{\mathrm{e}}_{n} \\ \ddot{\hat{\boldsymbol{w}}}^{\mathrm{e}}_{n+1} = \dfrac{1}{\beta\Delta t^2}\left(\hat{\boldsymbol{w}}^{\mathrm{e}}_{n+1} - \hat{\boldsymbol{w}}^{\mathrm{e}}_{n}\right) - \dfrac{1}{\beta\Delta t}\dot{\hat{\boldsymbol{w}}}^{\mathrm{e}}_{n} - \left(\dfrac{1}{2\beta}-1\right)\ddot{\hat{\boldsymbol{w}}}^{\mathrm{e}}_{n} \end{cases} \tag{7.1.7}$$

where γ and β are the Newmark parameters. When $\gamma \geqslant 1/2$ and $\beta \geqslant (\gamma+1/2)^2/4$, the Newmark scheme is unconditionally stable. Substituting Eqs. (7.1.6), (7.1.7) into a spatially discretized finite element equation (7.1.5), we have

$$\begin{cases} \boldsymbol{\Psi}^{\mathrm{s}}_{n+1}\left(\hat{\boldsymbol{u}}^{\mathrm{e}}_{n+1}, \hat{\boldsymbol{w}}^{\mathrm{e}}_{n+1}\right) = \boldsymbol{P}\left(\hat{\boldsymbol{u}}^{\mathrm{e}}_{n+1}\right) + \left[\boldsymbol{K}'_{\mathrm{ss}} + \dfrac{1}{\beta\Delta t^2}\boldsymbol{M}_{\mathrm{ss}}\right]\hat{\boldsymbol{u}}^{\mathrm{e}}_{n+1} \\ \qquad + \left[\boldsymbol{K}_{\mathrm{sw}} + \dfrac{1}{\beta\Delta t^2}\boldsymbol{M}_{\mathrm{sw}}\right]\hat{\boldsymbol{w}}^{\mathrm{e}}_{n+1} - \boldsymbol{F}^{\mathrm{s}}_{n+1} = \boldsymbol{0} \\ \boldsymbol{\Psi}^{\mathrm{w}}_{n+1}\left(\hat{\boldsymbol{u}}^{\mathrm{e}}_{n+1}, \hat{\boldsymbol{w}}^{\mathrm{e}}_{n+1}\right) = \left[\boldsymbol{K}_{\mathrm{ws}} + \dfrac{1}{\beta\Delta t^2}\boldsymbol{M}_{\mathrm{ws}}\right]\hat{\boldsymbol{u}}^{\mathrm{e}}_{n+1} \\ \qquad + \left[\boldsymbol{K}_{\mathrm{ww}} + \dfrac{1}{\beta\Delta t^2}\boldsymbol{M}_{\mathrm{ww}} + \dfrac{\gamma}{\beta\Delta t}\boldsymbol{C}_{\mathrm{ww}}\right]\hat{\boldsymbol{w}}^{\mathrm{e}}_{n+1} - \boldsymbol{F}^{\mathrm{w}}_{n+1} = \boldsymbol{0} \end{cases} \tag{7.1.8}$$

where

$$\begin{cases} \boldsymbol{F}^{\mathrm{s}}_{n+1} = \boldsymbol{f}^{\mathrm{s}}_{n+1} + \boldsymbol{M}_{\mathrm{ss}}\left[\dfrac{1}{\beta\Delta t^2}\hat{\boldsymbol{u}}^{\mathrm{e}}_n + \dfrac{1}{\beta\Delta t}\dot{\hat{\boldsymbol{u}}}^{\mathrm{e}}_n + \left(\dfrac{1}{2\beta}-1\right)\ddot{\hat{\boldsymbol{u}}}^{\mathrm{e}}_n\right] \\ \qquad + \boldsymbol{M}_{\mathrm{sw}}\left[\dfrac{1}{\beta\Delta t^2}\hat{\boldsymbol{w}}^{\mathrm{e}}_n + \dfrac{1}{\beta\Delta t}\dot{\hat{\boldsymbol{w}}}^{\mathrm{e}}_n + \left(\dfrac{1}{2\beta}-1\right)\ddot{\hat{\boldsymbol{w}}}^{\mathrm{e}}_n\right] \\ \boldsymbol{F}^{\mathrm{w}}_{n+1} = \boldsymbol{f}^{\mathrm{w}}_{n+1} + \boldsymbol{M}_{\mathrm{ws}}\left[\dfrac{1}{\beta\Delta t^2}\hat{\boldsymbol{u}}^{\mathrm{e}}_n + \dfrac{1}{\beta\Delta t}\dot{\hat{\boldsymbol{u}}}^{\mathrm{e}}_n + \left(\dfrac{1}{2\beta}-1\right)\ddot{\hat{\boldsymbol{u}}}^{\mathrm{e}}_n\right] \\ \qquad + \boldsymbol{M}_{\mathrm{ww}}\left[\dfrac{1}{\beta\Delta t^2}\hat{\boldsymbol{w}}^{\mathrm{e}}_n + \dfrac{1}{\beta\Delta t}\dot{\hat{\boldsymbol{w}}}^{\mathrm{e}}_n + \left(\dfrac{1}{2\beta}-1\right)\ddot{\hat{\boldsymbol{w}}}^{\mathrm{e}}_n\right] \\ \qquad + \boldsymbol{C}_{\mathrm{ww}}\left[\dfrac{\gamma}{\beta\Delta t}\hat{\boldsymbol{w}}^{\mathrm{e}}_n - \left(1-\dfrac{\gamma}{\beta}\right)\dot{\hat{\boldsymbol{w}}}^{\mathrm{e}}_n - \left(1-\dfrac{\gamma}{2\beta}\right)\Delta t\ddot{\hat{\boldsymbol{w}}}^{\mathrm{e}}_n\right] \\ \boldsymbol{P}\left(\hat{\boldsymbol{u}}^{\mathrm{e}}_{n+1}\right) = \displaystyle\int_\Omega \boldsymbol{B}^{\mathrm{T}}\boldsymbol{\sigma}''_{n+1}\mathrm{d}\Omega = \int_\Omega \boldsymbol{B}^{\mathrm{T}}\Delta\boldsymbol{\sigma}''_n\mathrm{d}\Omega + \boldsymbol{P}\left(\hat{\boldsymbol{u}}^{\mathrm{e}}_n\right) \end{cases} \tag{7.1.9}$$

For soil skeleton associated with an elasto-plastic behavior, the constitutive integration term $\boldsymbol{P}$ is nonlinear and dependent on the stress status and history. The resulting nonlinear equations need to be solved iteratively using the Newton-Raphson (N-R) procedure. Suppose the nodal displacement vectors $\hat{\boldsymbol{u}}^{em}_{n+1}$ and $\hat{\boldsymbol{w}}^{em}_{n+1}$ are known for the mth iteration at time ; then the Taylor expansion is performed for the governing Eq. (7.1.8) established for the $(m+1)$th iteration with respect to the unknowns $\hat{\boldsymbol{u}}^{em}_{n+1}$ and $\hat{\boldsymbol{w}}^{em}_{n+1}$. Only the linear terms of the expansion are kept; then:

$$\begin{aligned} \boldsymbol{\Psi}^{\mathrm{s}}_{n+1}\left(\hat{\boldsymbol{u}}^{\mathrm{e}(m+1)}_{n+1}, \hat{\boldsymbol{w}}^{\mathrm{e}(m+1)}_{n+1}\right) &\equiv \boldsymbol{\Psi}^{\mathrm{s}}_{n+1}\left(\hat{\boldsymbol{u}}^{em}_{n+1}, \hat{\boldsymbol{w}}^{em}_{n+1}\right) + \left.\frac{\partial \boldsymbol{\Psi}^{\mathrm{s}}}{\partial \hat{\boldsymbol{u}}^{\mathrm{e}}}\right|^m_{n+1}\Delta\hat{\boldsymbol{u}}^{em}_n + \left.\frac{\partial \boldsymbol{\Psi}^{\mathrm{s}}}{\partial \hat{\boldsymbol{w}}^{\mathrm{e}}}\right|^m_{n+1}\Delta\hat{\boldsymbol{w}}^{em}_n \\ \boldsymbol{\Psi}^{\mathrm{w}}_{n+1}\left(\hat{\boldsymbol{u}}^{\mathrm{e}(m+1)}_{n+1}, \hat{\boldsymbol{w}}^{\mathrm{e}(m+1)}_{n+1}\right) &\equiv \boldsymbol{\Psi}^{\mathrm{w}}_{n+1}\left(\hat{\boldsymbol{u}}^{em}_{n+1}, \hat{\boldsymbol{w}}^{em}_{n+1}\right) + \left.\frac{\partial \boldsymbol{\Psi}^{\mathrm{w}}}{\partial \hat{\boldsymbol{u}}^{\mathrm{e}}}\right|^m_{n+1}\Delta\hat{\boldsymbol{u}}^{em}_n + \left.\frac{\partial \boldsymbol{\Psi}^{\mathrm{w}}}{\partial \hat{\boldsymbol{w}}^{\mathrm{e}}}\right|^m_{n+1}\Delta\hat{\boldsymbol{w}}^{em}_n \end{aligned} \tag{7.1.10}$$

and

$$\left.\frac{\partial \boldsymbol{P}}{\partial \hat{\boldsymbol{u}}^{\mathrm{e}}}\right|^m_{n+1} = \left.\boldsymbol{K}^{\mathrm{T}}_{\mathrm{ss}}\right|^m_{n+1} \tag{7.1.11}$$

where $\boldsymbol{K}^{\mathrm{T}}_{\mathrm{ss}}$ is the tangential stiffness matrix of the soil skeleton; $\Delta\hat{\boldsymbol{u}}^{em}_n$ and $\Delta\hat{\boldsymbol{w}}^{em}_n$ denote the incremental displacement vectors for the soil skeleton and the pore fluid, respectively, at the $(m+1)$th N-R iteration.

For the incremental effective stress vector $\Delta\boldsymbol{\sigma}''_n|^m$ at the $(m+1)$th iteration, it can be related to the incremental strain using the updated elasto-plastic constitutive matrix of the soil skeleton:

$$\Delta\boldsymbol{\sigma}''_n\big|^m = \left(\boldsymbol{D} - \boldsymbol{D}_{\mathrm{p}}\big|^m_{n+1}\right)\Delta\boldsymbol{\varepsilon}_n\big|^m \tag{7.1.12}$$

thus

$$\int_{\Omega} \boldsymbol{B}^{\mathrm{T}} \Delta \boldsymbol{\sigma}_n''\big|^m \mathrm{d}\Omega = \int_{\Omega} \boldsymbol{B}^{\mathrm{T}} \left(\boldsymbol{D} - \boldsymbol{D}_{\mathrm{p}}\big|_{n+1}^m \right) \boldsymbol{B} \, \mathrm{d}\Omega \Delta \hat{\boldsymbol{u}}_n^{\mathrm{e}m} = \boldsymbol{K}_{\mathrm{ss}}^{\mathrm{T}}\big|_{n+1}^m \Delta \hat{\boldsymbol{u}}_n^{\mathrm{e}m} \tag{7.1.13}$$

After substituting Eq. (7.1.13) into Eq. (7.1.10), the original nonlinear governing equations can be linearized for the current (m + 1)th iteration as

$$\begin{bmatrix} \boldsymbol{K}_{\mathrm{ss}}^{\mathrm{T}}\big|_{n+1}^m + \boldsymbol{K}_{\mathrm{ss}}' + \dfrac{1}{\beta \Delta t^2} \boldsymbol{M}_{\mathrm{ss}} & \boldsymbol{K}_{\mathrm{sw}} + \dfrac{1}{\beta \Delta t^2} \boldsymbol{M}_{\mathrm{sw}} \\ \boldsymbol{K}_{\mathrm{ws}} + \dfrac{1}{\beta \Delta t^2} \boldsymbol{M}_{\mathrm{ws}} & \boldsymbol{K}_{\mathrm{ww}} + \dfrac{1}{\beta \Delta t^2} \boldsymbol{M}_{\mathrm{ww}} + \dfrac{\gamma}{\beta \Delta t} \boldsymbol{C}_{\mathrm{ww}} \end{bmatrix} \begin{Bmatrix} \Delta \hat{\boldsymbol{u}}_n^{\mathrm{e}m} \\ \Delta \hat{\boldsymbol{w}}_n^{\mathrm{e}m} \end{Bmatrix} = \begin{Bmatrix} -\boldsymbol{\Psi}_{n+1}^{\mathrm{s}} \left(\hat{\boldsymbol{u}}_{n+1}^{\mathrm{e}m}, \hat{\boldsymbol{w}}_{n+1}^{\mathrm{e}m} \right) \\ -\boldsymbol{\Psi}_{n+1}^{\mathrm{w}} \left(\hat{\boldsymbol{u}}_{n+1}^{\mathrm{e}m}, \hat{\boldsymbol{w}}_{n+1}^{\mathrm{e}m} \right) \end{Bmatrix} \tag{7.1.14}$$

where the coefficient matrix in brackets is the Jacobian matrix of the saturated soil element based on the $\boldsymbol{u}$-$\boldsymbol{w}$ formulation. It is symmetric since we have adopted the same shape function for the two phases. The updated displacement vectors of the soil skeleton and pore fluid for the (m + 1)th iteration at time (n+1)Δt is then obtained by solving Eq. (7.1.14):

$$\begin{cases} \hat{\boldsymbol{u}}_{n+1}^{\mathrm{e}(m+1)} = \hat{\boldsymbol{u}}_{n+1}^{\mathrm{e}m} + \Delta \hat{\boldsymbol{u}}_n^{\mathrm{e}m} \\ \hat{\boldsymbol{w}}_{n+1}^{\mathrm{e}(m+1)} = \hat{\boldsymbol{w}}_{n+1}^{\mathrm{e}m} + \Delta \hat{\boldsymbol{w}}_n^{\mathrm{e}m} \end{cases} \tag{7.1.15}$$

At the beginning of the iteration we can set $\boldsymbol{u}_{n+1}^{\mathrm{e}0} = \boldsymbol{u}_n^{\mathrm{e}}$. Suppose the convergence is reached after repeating the steps in Eqs. (7.1.10)–(7.1.14) for m_0 iterations; then the unknown nodal displacement vectors for the soil skeleton and the pore fluid at time are solved as

$$\begin{cases} \hat{\boldsymbol{u}}_{n+1}^{\mathrm{e}} = \hat{\boldsymbol{u}}_n^{\mathrm{e}} + \displaystyle\sum_{m=1}^{m_0} \Delta \hat{\boldsymbol{u}}_n^{\mathrm{e}m} \\ \hat{\boldsymbol{w}}_{n+1}^{\mathrm{e}} = \hat{\boldsymbol{w}}_n^{\mathrm{e}} + \displaystyle\sum_{m=1}^{m_0} \Delta \hat{\boldsymbol{w}}_n^{\mathrm{e}m} \end{cases} \tag{7.1.16}$$

The nodal velocity and acceleration vectors at time (n+1)Δt can be obtained by substituting Eq. (7.1.16) into Eqs. (7.1.6), (7.1.7) respectively for the soil skeleton and the pore fluid. Eq. (7.1.14) is the linearized governing equation for the saturated soil element involving an elasto-plastic soil skeleton. It was noted that a single N-R iteration solves the equation exactly in the linear case.

7.1.3 Verification of the Saturated Soil Element

Based on the derived finite element equations given in Section 7.1.2, we have developed a 2D/3D saturated soil element that was incorporated into an in-house finite code Train Saturated Soil Vibration Analysis Program (TSSVP). To verify its accuracy, firstly we established a saturated soil column and used it to study the 1D consolidation and wave-propagation problems. The results obtained from the calculation were compared with the analytical solutions available to the 1D problem. Then, we built a 2D plain strain model to study the dynamic responses of the soil under a dynamic load input, where saturated soil is reduced to an elastic soil by constraining the degrees-of-freedom (dof) associated with the pore water. The results obtained were verified with published numerical results, since no analytical solution was available for a 2D problem. A linear behavior was assumed for the saturated soil in all the following numerical experiments.

One-Dimensional 1D Consolidation of a Soil Column

A saturated soil column, with the column height h_0 and a surcharge loading p_0 applied to the top, was established and discretized using 100 equally spaced eight-node brick soil elements along the height (i.e., the vertical direction). The movements of both the soil skeleton and the pore water were constrained in the horizontal directions to mimic the 1D condition. The bottom of the column is impervious, while, the top of the column is pervious. The computational results were verified with the analytical Terzaghi solution for the single-side drainage consolidation, which can be found in many soil mechanics textbooks and the solution has been omitted here for brevity. The same material parameters of the saturated soil as those given by Jeremic et al.[53] were chosen, as shown in Table 7.1.1. It was noted that the Terzaghi solution assumed an incompressible behavior both for the soil grain and the pore water. Thus, the moduli of the soil grain and the pore water (K_s and K_f) in Table 7.1.1 take large values.

Table 7.1.1 Material parameters of the saturated soil in 1D consolidation simulation

Parameter	Value
Soil skeleton elastic modulus, E	2.0×10^7 kN/m^2
Soil skeleton Poisson's ratio, v	0.2
Soil skeleton mass density, ρ_s	2.0×10^3 kg/m^3
Pore water mass density, ρ_f	1.0×10^3 kg/m^3
Additional mass density, ρ_a	0 kg/m^3
Porosity, n	0.4
Darcy permeability, k_D	1.0×10^{-7} m/s
Soil grain modulus, K_s	1.0×10^{20} kN/m^2
Pore water modulus, K_f	2.2×10^9 kN/m^2
Gravity acceleration, g	9.8 m/s^2

The comparison of the pore water pressure results between the computational results and the analytical results is presented in Fig. 7.1.1. Its abscissa is the nondimensionalization of the excess pore water pressure p with respect to the surcharge loading amplitude p_0. And its ordinate is the nondimensionalization of the height h with respect to the column depth h_0. Numerical damping is introduced into the computation by setting the Newmark parameters, and so that the oscillations in the solutions caused by the sudden application of the surcharging load could be removed. It can be seen from Fig. 7.1.1 that the computational solutions agree very well with the Terzaghi solutions, which verifies the correctness of the developed saturated soil element based on the ***u*-*w*** formulation.

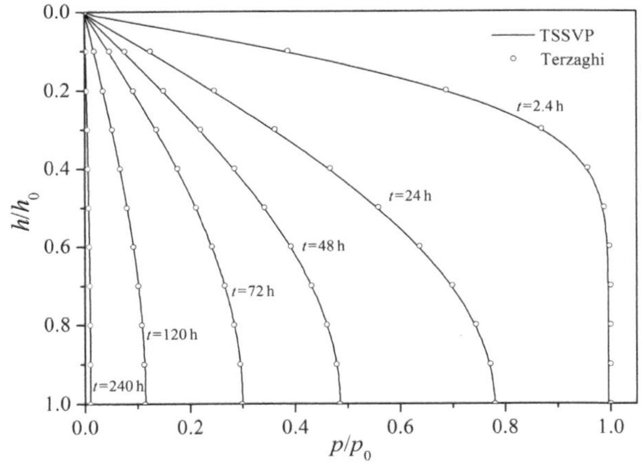

Fig. 7.1.1 Evolution of excess pore water pressure: a comparison between the TSSVP computational results and the analytical solution by Terzaghi

1D Wave Propagation in a Soil Column

Typically, there are three types of body waves, i.e., P1, P2, and S waves, existing in the saturated poroelastic medium. Due to the viscous coupling between the pore water and the soil skeleton, the three body waves are dispersive. To verify the accuracy of the developed saturated soil element in simulating the propagation of dispersive waves, a saturated soil column was established to study the 1D propagation of the longitudinal waves. The computational results were compared to the analytical results given by Gajo[51]. The column, with a width of 1 cm (x axis) and a height of 50 cm (y axis), was discretized using 5000 equally spaced four-node quadrilateral elements along the y axis. The movements of both the soil-skeleton and pore-water phases in the x direction were constrained to mimic 1D longitudinal wave propagation. A Heaviside step function with an amplitude of 0.001 cm was applied to the top of the column to achieve the soil-skeleton displacement u_y. The top of the column is impervious, i.e., w_y=0. At the base of the column, both u_y and w_y were constrained. The material parameters of the saturated soil were taken as those given by Gajo[83] and are shown in Table 7.1.2.

The displacement responses were recorded for an observation point located 1 cm below the top of the column. The total analysis time was set to 15 μs with a time step length of 0.01 μs. During the analysis time, the reflected wave from the bottom of the column did not arrive at the observation point, which implies that the boundary conditions at the base would not affect the responses at the observation point. The comparison on the displacement responses at the observation point between the computational solutions and the analytical solutions given by Gajo is present in Fig. 7.1.2.

Table 7.1.2 Material parameters of saturated soil in 1D wave propagation simulation

Parameter	Value
Soil skeleton elastic modulus, E	1.2×10^6 kN/m^2
Soil skeleton Poisson's ratio, v	0.3
Soil skeleton mass density, ρ_s	2.7×10^3 kg/m^3
Pore water mass density, ρ_f	1.0×10^3 kg/m^3
Additional mass density, ρ_a	0 kg/m^3
Porosity, n	0.4
Darcy permeability, k_D	1.0×10^{-9}, 1.0×10^{-5}, 1.0×10^{-1} m/s
Soil grain modulus, K_s	3.6×10^7 kN/m^2
Pore water modulus, K_f	2.17×10^6 kN/m^2
Gravity acceleration, g	9.8 m/s^2

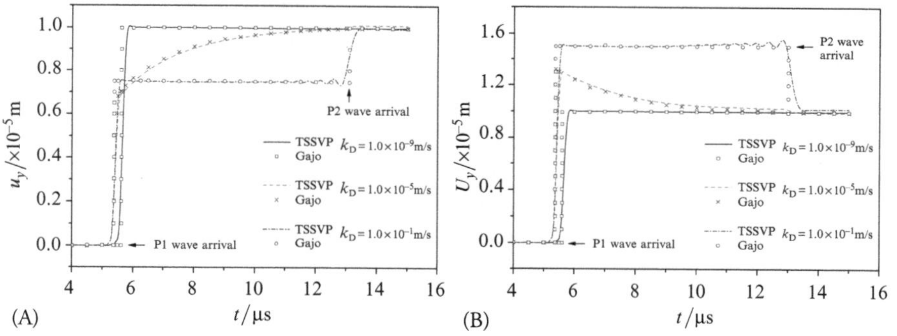

Fig. 7.1.2 Comparisons of the displacement histories at the observation point between computational results by TSSVP and analytical solutions given by Gajo. (A) u_y and (B) $U_y=u_y+w_y/n$

From Fig. 7.1.2 one can see that the computational results agree very well with the analytical solutions for all three permeabilities, and the simulation successfully reproduced the arrival time of the P1 and P2 waves. When the soil permeability was low (k_D=1.0×10^{-9} m/s), the viscous coupling between the soil-skeleton and the pore-water phases was strong, which resulted in a heavily damped P2 wave. Its contribution to the displacement responses at the observation point was negligible. For a high permeability (k_D=1.0×10^{-1} m/s), the viscous coupling is low and the P2 wave is a propagating wave. Thus its contribution is clearly visible in Fig. 7.1.2.

Plain-Strain Waves in a Saturated Square Soil Medium

A schematic view of the plain-strain rectangular domain occupied by a saturated soil medium is shown in Fig. 7.1.3A. A uniformly distributed edge loading with a width of 1 m (Fig. 7.1.3B) and lasting for T=0.16 s was applied vertically to the top left corner of the soil. The soil domain was spatially

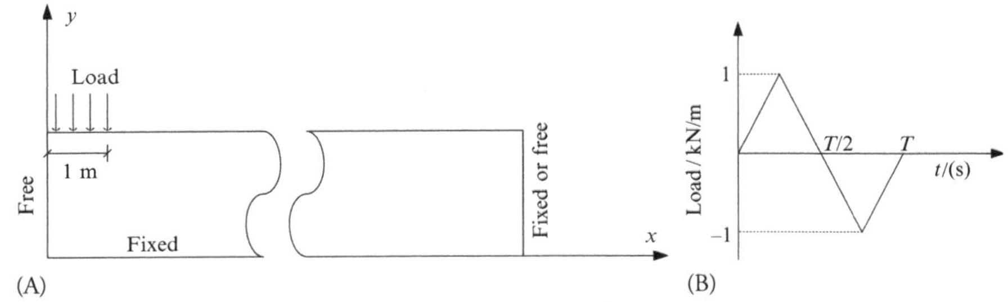

Fig. 7.1.3 Schematic view of a saturated soil medium and loads: (A) problem geometry and (B) distributed loading.

discretized by nine-node quadrilateral elements with a uniform size of 1m×1m. By constraining the degree of freedom associated with the pore water (i.e., w_x=w_y=0) and setting the Biot parameters α=M=0, the saturated soil element could be reduced to an elastic soil element. The displacement responses at the observation point A, which was located at the middle point of the top surface, was recorded and compared to the computational results given by Ju and Wang[58]. The soil base was constrained and its left side was free.

Fig. 7.1.4A and B present the comparisons of the horizontal displacement u_x for the right side of the soil being set free and constrained, respectively. The material parameters of the elastic soil that was reduced from the poroelastic soil were chosen as those given by Ju and Wang[58]: E=100 MPa, ν=0.3, and ρ=2000 kg/m^3. It can be seen from Fig. 7.1.4 that the TSSVP solutions agree very well with those from Ju and Wang[58], which demonstrates the accuracy of the developed saturated soil element in simulating the 2D wave propagations.

7.2 Absorbing Boundary Condition for the Saturated Soil Element

Due to the semiinfinite nature of the soil, the dynamic analysis of the soil via direct methods, such as finite element and finite difference, requires artificial boundaries to make the computational domain finite. The artificial boundary conditions should allow the wave motion to pass through and generate few or no reflections back into the finite computational domain. Such boundary conditions can be absorbing, transmitting, nonreflecting or silent boundary conditions. Absorbing boundary conditions (ABCs), which are local both in time and space, have attracted a lot of attention because they can be easily implemented in a finite element or finite difference analysis and can be employed to study nonlinear problems using time-step integration techniques.

Several well-known local ABCs have been proposed to transmit elastic waves in an elastic medium, such as the viscous boundary of Lysmer and Kuhlemeyer[59], the viscous-spring boundary of Deeks and Randolph[60], the para-axial boundary of Engquist and Majda[61] and Clayton and Engquist[62], the superposition boundary of Smith[63], the multidirectional boundary of Higdon[64], the high-order nonreflecting boundary condition of Dan Givoli and Beny Neta[65], the infinite element of Bettess and Zienkiewicz[66], the multitransmitting formula (MTF) of Liao and Wong[67], and the optimal absorbing boundary by Peng and Toksoz[68]. Among these, the MTF and the optimal absorbing boundary stand out since they are formulated on a discrete finite element/difference grid and are easy to implement in a finite element/difference analysis. The optimal ABC is constructed by properly locating the zeroes and poles of the reflection coefficients. However, the boundary coefficients are frequency dependent, which hinders their application in time-dependent problems.

The MTF is established with the aim of simulating the general transmitting process of the apparent plane wave propagating along a line normal to the artificial boundary. The MTF involves only 1D stencils, so it can be applied directly to the 2D and 3D cases and, particularly, no special treatment is needed at the interface of the material stratification and at the corners of the computational grid. Ideally speaking, each outgoing plane wave requires an apparent velocity along the line normal to the boundary. However, the incident directions of the waves are not available in advance and the physical velocities are different for each elastic wave type (e.g., S and P waves), which results in different apparent velocities along the line normal to the boundary. To overcome this, the developers of the MTF, Liao and Wong[67], introduced a common artificial wave velocity c_a to replace all the apparent velocities to formulate the MTF in a compact form; they suggested that c_a be the lowest physical wave velocity of the medium. By introducing several artificial wave velocities instead of the single wave velocity c_a in the space-time extrapolation, Liao[69] generalized the proposed MTF to consist of a system of nonreflection boundary conditions. A similar generalization of the MTF was achieved by Wagner and Chew[70] after optimizing the absorption of waves at specific angles of wave incidence. However, the numerical simulations of Wagner and Chew[70] and Liao[71] suggest that the generalization has little effect on the solution error when compared to the original MTF (i.e., simulating the apparent wave propagation process by employing a single artificial wave velocity).

The ABCs developed for the elastic medium have been extended to a saturated poroelastic medium in the literature. According to Biot's theory, the interaction between the solid skeleton phase and the pore-fluid phase results in three types of dispersive body waves, namely the P1, P2, and S waves, which makes the derivation of absorbing boundaries more difficult. Based on the 1D seismic wave incidence, Zienkiewicz[72] derived a viscous boundary for the ***u-p*** formulation of Biot's theory. At the expense of spurious reflections for oblique incident waves, Degrande and De Roeck[73] developed a viscous boundary for the ***u-w*** formulation in the frequency domain, which is local in space while nonlocal in time. By neglecting the P2 wave at the boundary, Akiyoushi et al.[74] and Modaressi and Benzenati[75] proposed viscous boundaries for the ***u-w***, ***u-U*** and ***u-p*** formulations. The viscous boundary conditions may suffer from stability problems when dealing with low-frequency dynamic loads. The stability can be improved by adding a spring element to the viscous boundary conditions[76], which takes the geometric attenuations of the wave field into consideration. By taking the asymptotic decay of the field quantity into account, Nenning and Schanz[77] formulated an infinite element for unbounded saturated porous media, which has been shown to be sufficiently accurate. However, the shape functions of the infinite element are constructed in the Laplace domain, and thus the infinite elements are nonlocal in time. Zhang[78] applied Smith's superposition boundary to the ***u*-*U*** formulation of a 2D saturated soil medium; however, this boundary condition can be very expensive in a 3D analysis. Gajo et al.[79] formed two types of multidirectional boundaries for the ***u*-*U*** formulation at null and infinite permeability, respectively. A preliminary judgment is needed before the proper boundary type can be selected. Zeng et al.[80] extended the perfectly matched layer (PML), which was originally proposed by Berenger[81] as a material-ABC for electromagnetic waves, to simulate seismic wave propagation in poroelastic media. Recently, Li and Song[82] derived a high-order transmitting boundary for cylindrical elastic wave propagation in infinite saturated porous media governed by the ***u*-*p*** formulation of Biot's theory with an assumption of zero permeability. The essence of the proposed transmitting boundary condition is to associate a set of spring, dashpot and mass elements to the boundary nodes. It is worth noting that ABCs involving spring elements or infinite elements need to approximate the decay of the field quantity as a function of r (r is distance measured to the applied load). For the stationary load inputs, the distance r from the boundary node to the applied load can be easily measured, while r is hard to determine for moving load inputs such as in simulations of ground vibrations caused by high-speed trains.

As it deals only with the displacements at grid points, the MTF can be directly extended to the saturated porous medium without simplification to Biot's theory (e.g., the neglecting of the P2 wave, null or infinite permeability). In the following, the MTF, in a discretized form, was proposed for the finite element modeling of wave propagation in a saturated poroelastic medium. First of all, the reflection coefficients of the MTF, to an arbitrary order N, were analytically derived in the frequency domain for the incidence of P1, P2, and SV waves to the artificial boundary truncating the finite element grid. Parametric studies were performed with respect to the excitation frequency and the artificial wave velocity, to investigate their influence on the reflection coefficient of the MTF. Then a finite element code incorporating the second-order MTF was developed to investigate the 1D, 2D, and 3D transient wave propagations in the saturated medium that were generated by the displacement and force inputs. The 1D numerical analysis reveals that the computational evaluations of the reflection coefficients of the MTF agree with the predicted values in the theoretical investigations. Specifically, the absorbing effect of the MTF was computationally examined for wave propagations in a saturated poroelastic medium generated by the moving load.

7.2.1 Theory

A uniform finite-element grid in the x-z plane is shown in Fig. 7.2.1. The computational domain is $\{(x,z):z \leqslant 0\}$and it is occupied by the saturated poroelastic medium. The MTF is applied along the x axis (z=0) and the z axis is normal to the artificial boundary. The grid size is Δs and the time-step length is Δt.

MTF Formulation

For the boundary point o in Fig. 7.2.1, the MTF extrapolates its displacement at time $t=(n+1)\Delta t$ as a linear combination of the displacements at previous time steps along a straight line normal to the artificial boundary in the grid's interior. Let u_i ($j\Delta s$, $-k\Delta s$, $n\Delta t$) denote the soil skeleton displacements

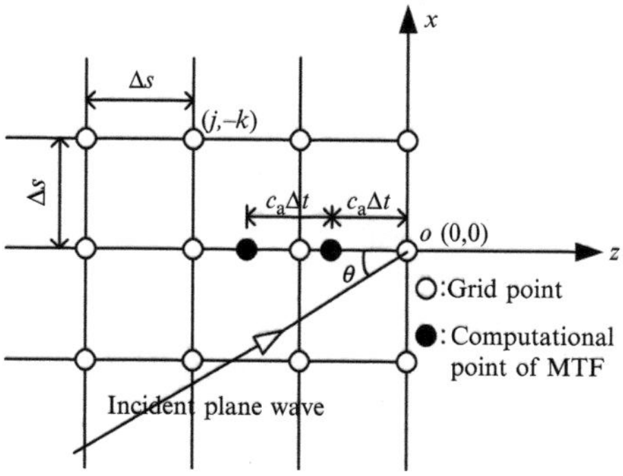

Fig. 7.2.1 Schematic view of the uniform finite element grid in the *x-z* plane and the MTF boundary condition applied at *z*=0

of the grid point at ($x=j\Delta s$, $z=-k\Delta s$) and at time $t=n\Delta t$ ($i=x$, z; $j=0$, ±1, ±2,…; k, n=0,1,2,…). Similarly, w_i ($j\Delta s$, $-k\Delta s$, $n\Delta t$) denotes the average pore fluid displacement relative to the soil skeleton at the grid point (j, k) and at time $n\Delta t$. The MTF is used to transmit displacements of the soil skeleton and the pore fluid at point o respectively:

$$\begin{cases} u_i(0,0,(n+1)\Delta t)=\sum_{k=1}^{N}(-1)^{k+1}C_k^N u_i(0,-kc_a\Delta t,(n+1-k)\Delta t) \\ w_i(0,0,(n+1)\Delta t)=\sum_{k=1}^{N}(-1)^{k+1}C_k^N w_i(0,-kc_a\Delta t,(n+1-k)\Delta t) \end{cases} \quad (i=x,z) \tag{7.2.1}$$

where c_a is an artificial wave velocity representing the apparent velocities of waves propagating along the z axis; $C_k^N=N!/[k!(N-k)!]$ is the binomial coefficient and N is the order of the boundary condition.

The computational points in Eq. (7.2.1) are located at $z=-kc_a\Delta t$, which do not generally coincide with the grid points in Fig. 7.2.1. The quadratic interpolation scheme was employed to express the displacement at the computational points in terms of those at the grid points as

$$A\phi(x,z,t)=\left(T_{11}+T_{12}Z_z^{-1}+T_{13}Z_z^{-2}\right)\phi(x,z,t)\approx\phi(x,z-c_a\Delta t,t) \tag{7.2.2}$$

where the operator A approximates a backward space shift of $c_a\Delta t$ and the interpolation coefficients are

$$\begin{cases} T_{11}=(2-s)(1-s)/2 \\ T_{12}=s(2-s) \\ T_{13}=s(s-1)/2 \\ s=c_a\Delta t/\Delta x \end{cases} \tag{7.2.3}$$

The operator *Zz* is the forward space shift given as (correspondingly, the operator Z_z^{-1} represents the backward space shift)

$$Z_z\phi(x, z, t) = \phi(x, z + \Delta s, t) \tag{7.2.4}$$

Similarly, we introduced the operator Z_t as the forward time shift (and the operator Z_z^{-1} represents the backward time shift) as

$$Z_t\phi(x, z, t) = \phi(x, z, t + \Delta t) \tag{7.2.5}$$

where $\phi(x, z, t)$ is an arbitrary function.

Using the operators defined in Eqs. (7.2.2)–(7.2.5), the MTF boundary conditions expressed in terms of displacements at the grid points were derived for the soil-skeleton and pore-fluid phases, respectively as

$$\begin{cases} \left[1 - AZ_t^{-1}\right]^N u_i(0, 0, (n+1)\Delta t) = 0 \\ \left[1 - AZ_t^{-1}\right]^N w_i(0, 0, (n+1)\Delta t) = 0 \end{cases} \quad (i = x, z) \tag{7.2.6}$$

Reflection Coefficients of the MTF

The governing equations of a saturated poroelastic medium established by Biot were given in Eqs. (1.1.24), (1.1.25) in Chapter 1. For in-plane motions, the displacements u_i and w_i (i=x, z) can be decomposed in terms of the scalar wave potentials φ^a and ψ^a:

$$\begin{cases} u_x = \dfrac{\partial\varphi^s}{\partial x} - \dfrac{\partial\psi^s}{\partial z}; \quad w_x = \dfrac{\partial\varphi^f}{\partial x} - \dfrac{\partial\psi^f}{\partial z} \\ u_z = \dfrac{\partial\varphi^s}{\partial z} + \dfrac{\partial\psi^s}{\partial x}; \quad w_z = \dfrac{\partial\varphi^f}{\partial z} + \dfrac{\partial\psi^f}{\partial x} \end{cases} \tag{7.2.7}$$

where the superscripts "s" and "f" denote the soil skeleton and pore fluid phase, respectively; φ^a and ψ^a (a = "s" or "f") are potentials of the longitudinal and shear waves, respectively.

After substituting Eq. (7.2.7) into the governing equations of the saturated soil and transforming the resulting equations into the wavenumber (η) and frequency (ω) domain with respect to x and t by using the Fourier transform defined in Eq. (H.1), an eigenvalue problem in terms of the wave potentials can be established. The eigenvalue Eq. (H.2) reveals the existence of two longitudinal waves (P1 and P2 waves) and one shear wave (S wave) in the saturated poroelastic medium.

At the artificial boundary of the saturated medium, the wave fields are composed of the incident and reflected waves. Hence, the scalar wave potentials governing the longitudinal and shear waves are expressed as

$$\begin{cases} \tilde{\bar{\varphi}}^{\mathrm{a}}(\eta, z, \omega) = \sum_{j=1}^{2}\left[P_j^{\mathrm{aI}} \mathrm{e}^{-\mathrm{i}k_{j\mathrm{pz}}^{\mathrm{I}} z} + P_j^{\mathrm{aR}} \mathrm{e}^{\mathrm{i}k_{j\mathrm{pz}}^{\mathrm{R}} z}\right] \\ \tilde{\bar{\psi}}^{\mathrm{a}}(\eta, z, \omega) = S^{\mathrm{aI}} \mathrm{e}^{-\mathrm{i}k_{\mathrm{sz}}^{\mathrm{I}} z} + S^{\mathrm{aR}} \mathrm{e}^{\mathrm{i}k_{\mathrm{sz}}^{\mathrm{R}} z} \end{cases} \tag{7.2.8}$$

where the superscripts "~" and "–" denote a variable in the frequency and wavenumber domains, respectively; the superscripts "I" and "R" refer to the incident and reflected waves; the subscripts j=1 and 2 attribute variables to the P1 and P2 wave, respectively; P^{a} and S^{a} are the unknown potential amplitudes; k_{pz} and k_{sz} are the vertical components (along the z axis) of the compressional wavenumber $\boldsymbol{k}_{\mathrm{p}}(\eta, k_{\mathrm{pz}})$ and the shear wavenumber $\boldsymbol{k}_{\mathrm{s}}(\eta, k_{\mathrm{sz}})$, respectively, which can be solved using Eq. (H.2).

Transform the discrete MTF formulation in Eq. (7.2.6) and the wave potential decomposition Eq. (7.2.7) into wavenumber (η) and frequency (ω) domain with respect to x and t by using the Fourier transformation:

$$\begin{cases} \left[1 - A\mathrm{e}^{-\mathrm{i}\omega\Delta t}\right]^N \tilde{\bar{u}}_i(\eta, 0, \omega)\mathrm{e}^{\mathrm{i}\omega\Delta t} = 0 \\ \left[1 - A\mathrm{e}^{-\mathrm{i}\omega\Delta t}\right]^N \tilde{\bar{w}}_i(\eta, 0, \omega)\mathrm{e}^{\mathrm{i}\omega\Delta t} = 0 \end{cases} \quad (i = x, z) \tag{7.2.9}$$

$$\begin{cases} \tilde{\bar{u}}_x = -\mathrm{i}\eta\tilde{\bar{\varphi}}^{\mathrm{s}} - \dfrac{\partial\tilde{\bar{\psi}}^{\mathrm{s}}}{\partial z}; \tilde{\bar{w}}_x = -\mathrm{i}\eta\tilde{\bar{\varphi}}^{\mathrm{f}} - \dfrac{\partial\tilde{\bar{\psi}}^{\mathrm{f}}}{\partial z} \\ \tilde{\bar{u}}_z = \dfrac{\partial\tilde{\bar{\varphi}}^{\mathrm{s}}}{\partial z} - \mathrm{i}\eta\tilde{\bar{\psi}}^{\mathrm{s}}; \tilde{\bar{w}}_z = \dfrac{\partial\tilde{\bar{\varphi}}^{\mathrm{f}}}{\partial z} - \mathrm{i}\eta\tilde{\bar{\psi}}^{\mathrm{f}} \end{cases} \tag{7.2.10}$$

Substituting Eq. (7.2.8) into Eq. (7.2.10), the soil-skeleton and pore-fluid relative displacements in the transformed domain can be readily expressed by the wave potential amplitudes. By substituting the obtained displacements $\tilde{\bar{u}}_i$ and $\tilde{\bar{w}}_i$ (i=x, z) into the transformed MTF formulation in Eq. (7.2.9), four equations in terms of six unknowns, i.e., S^{sI}, S^{sR}, P_j^{sI}, and P_j^{sR} (j=1, 2), are established. The unknown potential amplitudes of the pore fluid phase, i.e., S^{fI}, S^{fR}, P_j^{fI}, and P_j^{fR}, are dependent on these six unknowns through Eq. (H.5). When a particular wave (e.g., SV, P1, or P2 wave) impinges on the artificial boundary, reflected waves of all types arise. Taking the incidence of the SV wave as an example, the three reflected wave amplitudes, i.e., P_j^{sR} (j=1, 2) and S^{sR}, can be solved from the four MTF equations by setting P_j^{sI}=0 and S^{sI}=1. Amplitudes of the reflected P1, P2, and SV waves relative to the amplitude of the incident SV wave are denoted as $R_{\mathrm{ac}}=P_1^{\mathrm{sR}}/S^{\mathrm{sI}}$, $R_{\mathrm{bc}}=P_2^{\mathrm{sR}}/S^{\mathrm{sI}}$, and $R_{\mathrm{cc}}=S^{\mathrm{sR}}/S^{\mathrm{sI}}$, respectively. Similarly, for the incidence of the P1 wave, the relative amplitudes of the reflected P1, P2, and SV waves are denoted as R_{aa}, R_{ba}, and R_{ca} and for the incidence of the P2 wave, the relative amplitudes of the reflected P1, P2, and SV waves are denoted as R_{ab}, R_{bb}, and R_{cb}, respectively. The expressions of these relative amplitudes, which can also be termed reflection coefficients, are given as

SV wave incidence

$$\begin{cases} R_{\mathrm{ac}} = \dfrac{co_{2\mathrm{p}} - co_{\mathrm{s}}}{co_{2\mathrm{p}} - co_{1\mathrm{p}}} \times \dfrac{\left(k_{\mathrm{sz}}^{\mathrm{I}} + k_{\mathrm{sz}}^{\mathrm{R}}\right)\eta}{k_{1\mathrm{pz}}^{\mathrm{R}} k_{\mathrm{sz}}^{\mathrm{R}} + \eta^2} \times \dfrac{\left[1 - A\left(k_{\mathrm{sz}}^{\mathrm{I}}\right)\mathrm{e}^{-\mathrm{i}\omega\Delta t}\right]^N}{\left[1 - A\left(k_{1\mathrm{pz}}^{\mathrm{R}}\right)\mathrm{e}^{-\mathrm{i}\omega\Delta t}\right]^N} \\ R_{\mathrm{bc}} = \dfrac{co_{1\mathrm{p}} - co_{\mathrm{s}}}{co_{1\mathrm{p}} - co_{2\mathrm{p}}} \times \dfrac{\left(k_{\mathrm{sz}}^{\mathrm{I}} + k_{\mathrm{sz}}^{\mathrm{R}}\right)\eta}{k_{2\mathrm{pz}}^{\mathrm{R}} k_{\mathrm{sz}}^{\mathrm{R}} + \eta^2} \times \dfrac{\left[1 - A\left(k_{\mathrm{sz}}^{\mathrm{I}}\right)\mathrm{e}^{-\mathrm{i}\omega\Delta t}\right]^N}{\left[1 - A\left(k_{2\mathrm{pz}}^{\mathrm{R}}\right)\mathrm{e}^{-\mathrm{i}\omega\Delta t}\right]^N} \\ R_{\mathrm{cc}} = \dfrac{k_{\mathrm{sz}}^{\mathrm{I}} k_{1\mathrm{pz}}^{\mathrm{R}} - \eta^2}{k_{\mathrm{sz}}^{\mathrm{R}} k_{1\mathrm{pz}}^{\mathrm{R}} + \eta^2} \times \dfrac{\left[1 - A\left(k_{\mathrm{sz}}^{\mathrm{I}}\right)\mathrm{e}^{-\mathrm{i}\omega\Delta t}\right]^N}{\left[1 - A\left(k_{\mathrm{sz}}^{\mathrm{R}}\right)\mathrm{e}^{-\mathrm{i}\omega\Delta t}\right]^N} \end{cases} \tag{7.2.11}$$

P1 wave incidence

$$\begin{cases} R_{\mathrm{aa}} = \dfrac{k_{1\mathrm{pz}}^{\mathrm{I}} k_{\mathrm{sz}}^{\mathrm{R}} - \eta^2}{k_{1\mathrm{pz}}^{\mathrm{R}} k_{\mathrm{sz}}^{\mathrm{R}} + \eta^2} \times \dfrac{\left[1 - A\left(k_{1\mathrm{pz}}^{\mathrm{I}}\right)\mathrm{e}^{-\mathrm{i}\omega\Delta t}\right]^N}{\left[1 - A\left(k_{1\mathrm{pz}}^{\mathrm{R}}\right)\mathrm{e}^{-\mathrm{i}\omega\Delta t}\right]^N} \\ R_{\mathrm{ba}} = \dfrac{co_{1\mathrm{p}} - co_{\mathrm{s}}}{co_{2\mathrm{p}} - co_{\mathrm{s}}} \times \dfrac{k_{1\mathrm{pz}}^{\mathrm{I}} + k_{1\mathrm{pz}}^{\mathrm{R}}}{k_{2\mathrm{pz}}^{\mathrm{R}} - k_{1\mathrm{pz}}^{\mathrm{R}}} \times \dfrac{\left[1 - A\left(k_{1\mathrm{pz}}^{\mathrm{I}}\right)\mathrm{e}^{-\mathrm{i}\omega\Delta t}\right]^N}{\left[1 - A\left(k_{2\mathrm{pz}}^{\mathrm{R}}\right)\mathrm{e}^{-\mathrm{i}\omega\Delta t}\right]^N} \\ R_{\mathrm{ca}} = \dfrac{co_{1\mathrm{p}} - co_{2\mathrm{p}}}{co_{2\mathrm{p}} - co_{\mathrm{s}}} \times \dfrac{\left(k_{1\mathrm{pz}}^{\mathrm{I}} + k_{1\mathrm{pz}}^{\mathrm{R}}\right)\eta}{k_{\mathrm{sz}}^{\mathrm{R}} k_{1\mathrm{pz}}^{\mathrm{R}} + \eta^2} \times \dfrac{\left[1 - A\left(k_{1\mathrm{pz}}^{\mathrm{I}}\right)\mathrm{e}^{-\mathrm{i}\omega\Delta t}\right]^N}{\left[1 - A\left(k_{\mathrm{sz}}^{\mathrm{R}}\right)\mathrm{e}^{-\mathrm{i}\omega\Delta t}\right]^N} \end{cases} \tag{7.2.12}$$

P2 wave incidence

$$\begin{cases} R_{\mathrm{ab}} = \dfrac{co_{2\mathrm{p}} - co_{\mathrm{s}}}{co_{1\mathrm{p}} - co_{\mathrm{s}}} \times \dfrac{k_{2\mathrm{pz}}^{\mathrm{I}} + k_{2\mathrm{pz}}^{\mathrm{R}}}{k_{1\mathrm{pz}}^{\mathrm{R}} - k_{2\mathrm{pz}}^{\mathrm{R}}} \times \dfrac{\left[1 - A\left(k_{2\mathrm{pz}}^{\mathrm{I}}\right)\mathrm{e}^{-\mathrm{i}\omega\Delta t}\right]^N}{\left[1 - A\left(k_{1\mathrm{pz}}^{\mathrm{R}}\right)\mathrm{e}^{-\mathrm{i}\omega\Delta t}\right]^N} \\ R_{\mathrm{bb}} = \dfrac{k_{2\mathrm{pz}}^{\mathrm{I}} k_{\mathrm{sz}}^{\mathrm{R}} - \eta^2}{k_{2\mathrm{pz}}^{\mathrm{R}} k_{\mathrm{sz}}^{\mathrm{R}} + \eta^2} \times \dfrac{\left[1 - A\left(k_{2\mathrm{pz}}^{\mathrm{I}}\right)\mathrm{e}^{-\mathrm{i}\omega\Delta t}\right]^N}{\left[1 - A\left(k_{2\mathrm{pz}}^{\mathrm{R}}\right)\mathrm{e}^{-\mathrm{i}\omega\Delta t}\right]^N} \\ R_{\mathrm{cb}} = \dfrac{co_{1\mathrm{p}} - co_{2\mathrm{p}}}{co_{\mathrm{s}} - co_{1\mathrm{p}}} \times \dfrac{\left(k_{2\mathrm{pz}}^{\mathrm{I}} + k_{2\mathrm{pz}}^{\mathrm{R}}\right)\eta}{k_{\mathrm{sz}}^{\mathrm{R}} k_{2\mathrm{pz}}^{\mathrm{R}} + \eta^2} \times \dfrac{\left[1 - A\left(k_{2\mathrm{pz}}^{\mathrm{I}}\right)\mathrm{e}^{-\mathrm{i}\omega\Delta t}\right]^N}{\left[1 - A\left(k_{\mathrm{sz}}^{\mathrm{R}}\right)\mathrm{e}^{-\mathrm{i}\omega\Delta t}\right]^N} \end{cases} \tag{7.2.13}$$

where the coefficients co_{s} and $co_{j\mathrm{p}}$ (j=1, 2), which relate the potential amplitude of the pore-fluid phase to that of the soil-skeleton phase, are given in Eq. (H.6); the operator A, when applied to the incident and reflected P1, P2, and SV waves, is expressed as

$$\begin{cases} A\left(k_{jpz}^{\mathrm{I}}\right) = T_{11} + T_{12}\mathrm{e}^{\mathrm{i}k_{jpz}^{\mathrm{I}}\Delta s} + T_{13}\mathrm{e}^{\mathrm{i}2k_{jpz}^{\mathrm{I}}\Delta s} \\ A\left(k_{jpz}^{\mathrm{R}}\right) = T_{11} + T_{12}\mathrm{e}^{-\mathrm{i}k_{jpz}^{\mathrm{R}}\Delta s} + T_{13}\mathrm{e}^{-\mathrm{i}2k_{jpz}^{\mathrm{R}}\Delta s} \end{cases} \quad (j=1,2) \tag{7.2.14}$$

$$\begin{cases} A\left(k_{sz}^{\mathrm{I}}\right) = T_{11} + T_{12}\mathrm{e}^{\mathrm{i}k_{sz}^{\mathrm{I}}\Delta s} + T_{13}\mathrm{e}^{\mathrm{i}2k_{sz}^{\mathrm{I}}\Delta s} \\ A\left(k_{sz}^{\mathrm{R}}\right) = T_{11} + T_{12}\mathrm{e}^{-\mathrm{i}k_{sz}^{\mathrm{R}}\Delta s} + T_{13}\mathrm{e}^{-\mathrm{i}2k_{sz}^{\mathrm{R}}\Delta s} \end{cases} \tag{7.2.15}$$

By denoting the incidence angle of the SV, P1, and P2 waves as θ_{sv}^{I}, θ_{1p}^{I}, and θ_{2p}^{I}, respectively, the reflection angles θ_{sv}^{R}, θ_{1p}^{R}, and θ_{2p}^{R} at the artificial boundary can be determined by Snell's law:

$$k_s \sin\theta_{sv}^{\mathrm{R}}=k_{1p}\sin\theta_{1p}^{\mathrm{R}}=k_{2p}\sin\theta_{2p}^{\mathrm{R}}=\eta \tag{7.2.16}$$

where $\eta=k_s \sin\theta_{sv}^{\mathrm{I}}$, $k_{1p}\sin\theta_{1p}^{\mathrm{I}}$, and $k_{2p}\sin\theta_{2p}^{\mathrm{I}}$, respectively, for the incidence of SV, P1, and P2 waves. The vertical components (i.e., components along the z axis) of the wavenumbers of the incident SV, P1, and P2 waves can then be determined as $k_{sz}^{\mathrm{I}}=k_s\cos\theta_{sv}^{\mathrm{I}}$, $k_{1pz}^{\mathrm{I}}=k_{1p}\cos\theta_{1p}^{\mathrm{I}}$, and $k_{2pz}^{\mathrm{I}}=k_{2p}\cos\theta_{2p}^{\mathrm{I}}$, respectively. Correspondingly, the vertical components of the wavenumbers of the reflected SV, P1, and P2 waves are given as $k_{sz}^{\mathrm{R}}=k_s\cos\theta_{sv}^{\mathrm{R}}$, $k_{1pz}^{\mathrm{R}}=k_{1p}\cos\theta_{1p}^{\mathrm{R}}$, and $k_{2pz}^{\mathrm{R}}=k_{2p}\cos\theta_{2p}^{\mathrm{R}}$. Eqs. (7.2.11)–(7.2.13) constitute the analytical solutions of the reflection coefficients when the MTF is applied at boundaries of a finite element grid discretizing a saturated poroelastic medium.

Wave Reflection Analysis

It can be seen from Eqs. (7.2.11)–(7.2.13) that the terms of the reflection coefficients in square brackets are functions of three dimensionless parameters, i.e., $s=c_a\,\Delta t/\Delta s$, $\Delta t/T$, and $\Delta s/\lambda_{\min}$, which characterize the MTF formulation and the finite element grid. In the previous, $T=2\pi/\omega$ and $\lambda_{\min}=2\pi/k_{\max}$ denote the wave period and the minimum wave length of the body waves in the poroelastic medium at the frequency ω, respectively. Note that the wave velocities in a saturated medium under the fully undrained condition are the smallest; thus, the minimum wavelength can be computed as $\lambda_{\min}=\lambda_{s0}=2\pi C_{s0}/\omega$. Here C_{s0} is the velocity of the SV wave at the fully undrained limit and its expression is given in Eq. (H.3). The other terms in the expressions of the reflection coefficients are functions of the incident angle (i.e., θ_{sv}^{I} and θ_{jp}^{I}, $j=1$, 2) and the frequency component ω, which are not dependent on the computational grid. The frequency component ω can be expressed in a nondimensional form as $\chi=\omega/\omega_0$, where ω_0 is the characteristic frequency of the saturated poroelastic medium[73], and its expression is given in Eq. (H.2).

This section will analyze the effects of the dimensionless frequency ratio χ and the artificial wave velocity c_a on the reflection coefficients of the MTF for the incidence of SV, P1, and P2 waves. As the finite-element study of wave propagations requires at least 6–8 elements per minimum wavelength, the dimensionless grid size is taken to be $\Delta s/\lambda_{s0}=0.1$. The dimensionless time-step length $\Delta t/T$ is determined so that the dimensionless interpolation parameter $s=c_a\Delta t/\Delta s=0.5$ is reached. Material parameters of the water-saturated poroelastic medium are chosen as those given by Gajo et al.[83] The material parameters and the wave velocities at the two extreme conditions (i.e., the fully drained and fully undrained conditions given in Eq. (H.3) are presented in Table 7.2.1.

Table 7.2.1 Material parameters of saturated poroelastic medium and wave velocities at extreme conditions

Parameter	Value
Soil skeleton Lame constant, μ	4.62×10^8 N/m^2
Soil skeleton Lame constant, λ	6.92×10^8 N/m^2
Soil skeleton mass density, ρ_s	2.7×10^3 kg/m^3
Pore water mass density, ρ_f	1.0×10^3 kg/m^3
Tortuosity factor, a	1.0
Porosity, n	0.4
Biot's parameter, α	0.97
Biot's parameter, M	5.0×10^9 N/m^2
Darcy permeability, k_D	1.0×10^{-4} m/s
Gravity acceleration, g	9.81 m/s^2
Characteristic frequency, ω_0	4.89×10^4 rad/s
P1 wave velocity at FDC, C_{p1}	1849.7 m/s
P2 wave velocity at FDC, C_{p2}	763.1 m/s
SV wave velocity at FDC, C_s	533.7 m/s
P wave velocity at FUC, C_{p0}	1770.8 m/s
SV wave velocity at FUC, C_{s0}	478 m/s

FDC: fully drained condition. FUC: fully undrained condition.

Influences of the Dimensionless Frequency Ratio χ

In order to investigate the MTF behavior at high, intermediate and low viscous couplings between the solid-skeleton and pore-water phases, three values of χ=0.001, 0.1, and 10 were considered. The artificial wave velocity was chosen as $c_a=C_{s0}$ and the dimensionless time-step length was set at $\Delta t/T$=0.05. Then $s=C_{s0}\Delta t/\Delta s$=0.5 can be obtained. Fig. 7.2.2 presents the reflection coefficient of the second-order MTF (N=2) when the plane SV wave impinges on the artificial boundary.

For the whole range of the incident angle, the relative amplitudes R_{ac} and R_{bc} of the reflected P1 and P2 waves are both below 0.14, as shown in Fig. 7.2.2A and B. From Fig. 7.2.2C it is seen that R_{cc} is lower than 0.1 when the incident angle is below 50° and it goes up almost linearly to unity with the incident angle increasing to 90°. When χ increases from 0.001 to 10, R_{bc} is substantially increased from a value close to 0 to 0.12, while the influence of χ on R_{ac} and R_{cc} is limited. This phenomenon may be explained as the P2 wave in the saturated medium is more dispersive than the P1 and SV waves. In fact the P2 wave transfers from an evanescent wave ($V_{2p}\approx 0$) to a propagation wave ($V_{2p}\approx C_{p2}$) when χ increases from 0.001 to 10, where V_{2p} is the P2 wave velocity at the current frequency (Eq. (H.4)).

The relative amplitudes of the reflected P1, P2, and SV waves at the artificial boundary are presented in Figs. 7.2.3 and 7.2.4 respectively for the plane P1 and P2 wave incidences.

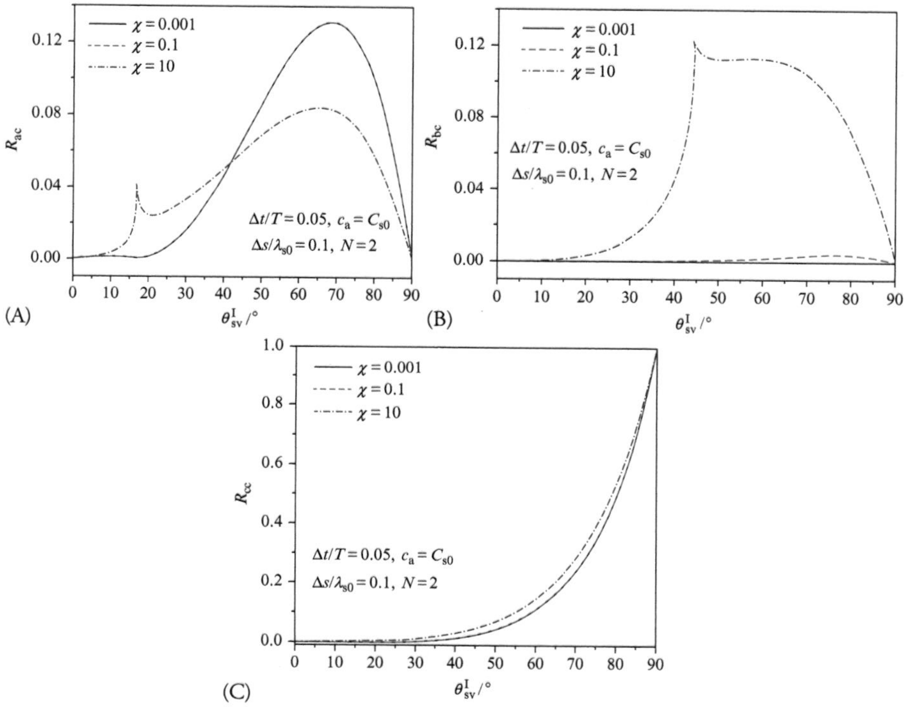

Fig. 7.2.2 Reflection coefficients of the second-order MTF for the plane SV wave incidence when χ=0.001, 0.1, and 10. (A) R_{ac}; (B) R_{bc}; and (C) R_{cc}

From Fig. 7.2.3A, it is observed that the relative amplitude R_{aa} of the reflected P1 wave is large (around 0.35) for the whole range of the P1-wave incident angle except at around 75°. The dip at the incident angle of 75° is accounted for by the fact that the numerator $k^{I}_{1pz}\ k^{R}_{sz}-\eta^2$ in R_{aa} is zero, which leads to

$$\sin^2\theta^{I}_{1p}=\frac{1}{\left(V_s^2/V_{1p}^2+1\right)} \tag{7.2.17}$$

where V_{1p} and V_s are the P1 and SV wave velocities at the current frequency (Eq. G.4), respectively. It is obtained from Eq. (7.2.17) that θ^{I}_{1p}=73.91°, which agrees with the observation in Fig. 7.2.3A. The relative amplitudes R_{ba} and R_{ca} of the reflected P2 and SV waves are below 0.3 and 0.18, respectively, as shown in Fig. 7.2.3B and C. The large R_{aa} is rendered because of the big difference between the P1-wave velocity V_{1p} and the artificial wave velocity c_a=C_{s0} adopted in the MTF formulation, i.e., $V_{1p} \approx 4C_{s0}$. For the normal incidence of the P1 wave (θ^{I}_{1p}=0°), it can be seen from Fig. 7.2.3A that R_{aa}=0.33 for $\chi \leq 0.1$ and R_{aa}=0.35 for χ=10. Fig. 7.2.3B shows that R_{ba} is significantly amplified when χ increases from 0.1 to 10, which is the same as the observation made in Fig. 7.2.2B.

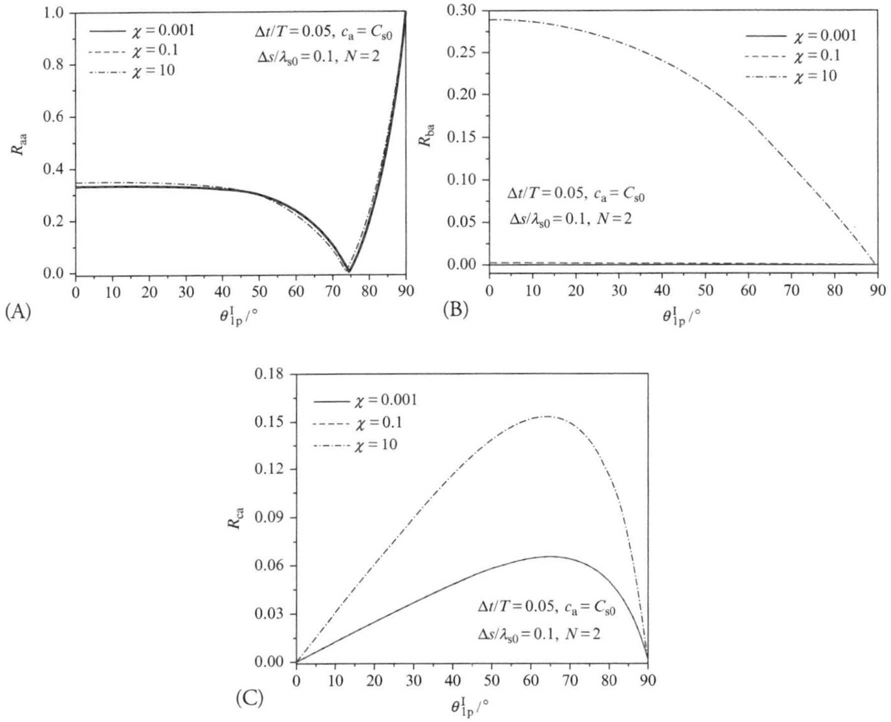

Fig. 7.2.3 Reflection coefficients of the second-order MTF for the plane P1 wave incidence when χ=10, 0.1, and 10. (A) R_{aa}; (B) R_{ba}; and (C) R_{ca}

When χ=0.001 and 0.1, the viscous coupling between the two phases of the saturated medium is dominant, which results in a nonpropagating P2 wave. Under this circumstance, the relative amplitudes of the reflected P1, P2, and SV waves at the artificial boundary will be significantly larger than unity when the plane P2 wave is enforced to strike the boundary. Thus, in Fig. 7.2.4 only the reflection coefficients are presented at χ=10. Although the P2 wave is of a propagation type at χ=10, it is heavily damped as the excitation frequency is high. Hence, for the artificial boundary located at a distance from the vibration source, the generated P2 wave may never arrive at the artificial boundary.

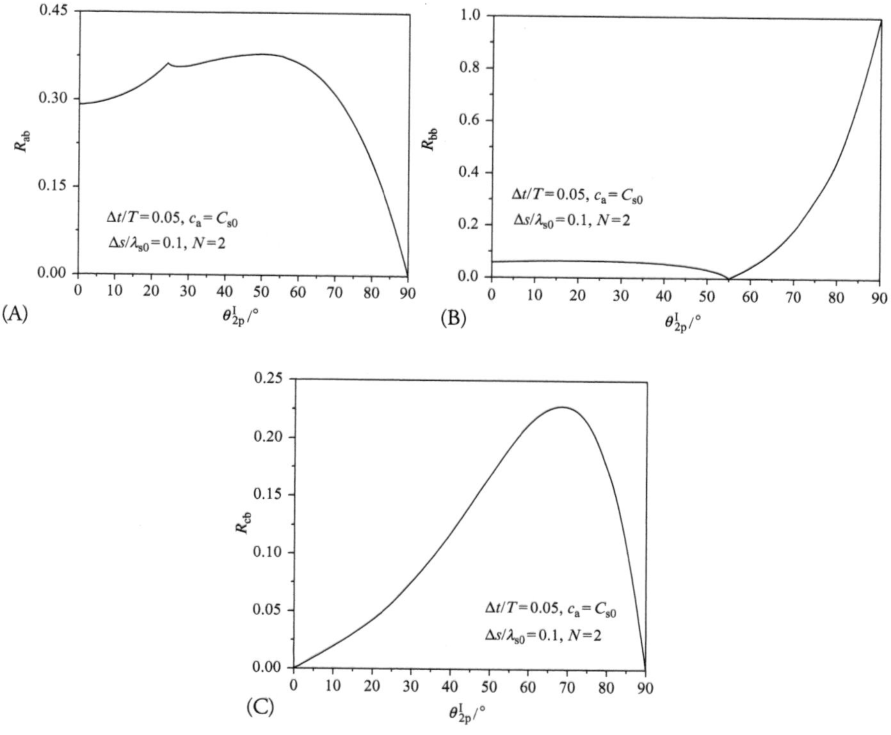

Fig. 7.2.4 Reflection coefficients of the second-order MTF for the plane P2 wave incidence when χ=10. (A) R_{ab}; (B) R_{bb}; and (C) R_{cb}

Influences of the Artificial Wave Velocity C_a

As three dispersive body waves (i.e., P1, P2, and SV waves) typically exist in the saturated poroelastic medium, we set velocities of the three different body waves at the extreme conditions, i.e., the shear wave velocity C_{s0} and the longitudinal wave velocity C_{p0} at the fully undrained condition and the second longitudinal wave velocity C_{p2} at the fully drained condition, respectively, to be the artificial wave velocity c_a in order to investigate their influence on the reflection coefficients. Fig. 7.2.5 compares the three different values of c_a when the dimensionless frequency ratio . The expressions of C_{s0}, C_{p0}, and C_{p2} can be found in Eq. (H.3) and their values are given in Table 7.2.1. The dimensionless time-step length $\Delta t/T$ is set to 0.05, 0.013, and 0.031 when c_a takes the value of C_{s0}, C_{p0}, and C_{p2}, respectively. Then $s=c_a \Delta t/\Delta s=0.5$ is ensured.

For the incidence of the P1 wave, it is evident from Fig. 7.2.5A that the reflection coefficient R_{aa} is substantially decreased when the artificial wave velocity c_a changes from the value of the shear or second longitudinal wave velocity (i.e., C_{s0} or C_{p2}) to a value close to the first longitudinal wave velocity (i.e., C_{p0}). A similar observation can be made in Fig. 7.2.5B and C when the incident angle is in the vicinity of zero (i.e., normal incident): R_{bb} and R_{cc} reach their lowest values when c_a equals C_{p2}

and C_{s0}, respectively. From the comparison it can be remarked that the reflection coefficients can be significantly reduced when the artificial velocity c_a is set close to the velocity of the incident wave, and no optimal choice of c_a exists to rend the reflection coefficients of all three types of body waves to their smallest simultaneously.

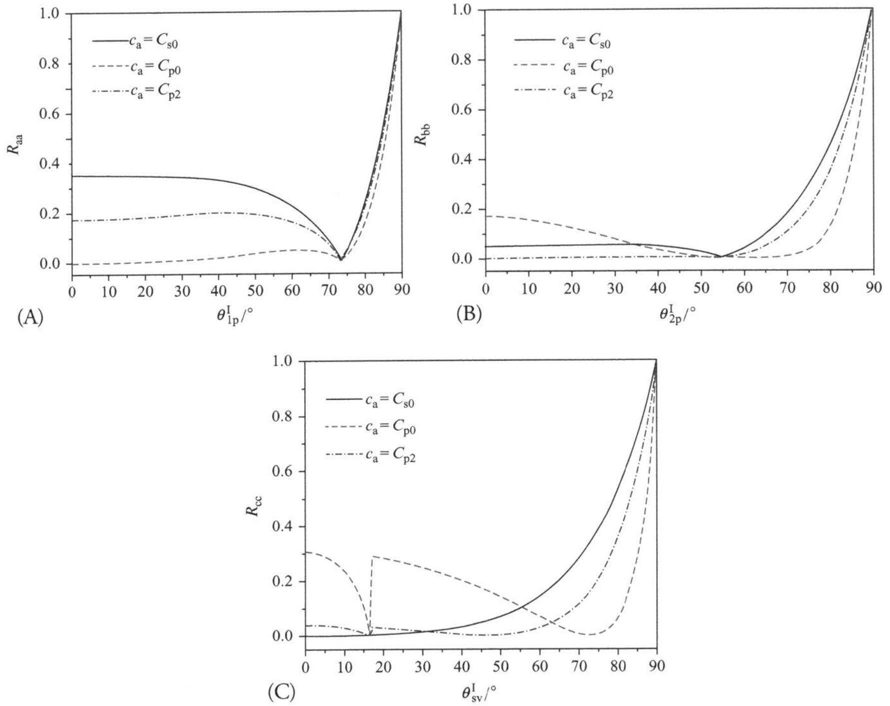

Fig. 7.2.5 Reflection coefficients of the second-order MTF for three different values of c_a at $\chi=10$. (A) R_{aa} for P1 wave incidence; (B) R_{bb} for P2 wave incidence; and (C) R_{cc} for SV wave incidence

The absorbing effect MTF for a particular wave type (e.g., SV wave) can be maximized only with some sacrifice of the absorption of the other two waves (e.g., P1 and P2 waves) if a single artificial wave velocity is adopted. However, we have no intention to generalize the MTF formulation for taking several different artificial wave velocities into account; the numerical experiments of Wagner and Chew[70] and Liao[71] demonstrated that such a generalization has little effect on the absorbing efficiency. Here we provide general guidelines on the choice of the single artificial wave velocity c_a: if the type and direction of the wave approaching the artificial boundary are unique, then one should set c_a to be the apparent velocity (i.e., $c_a/\cos\theta$, where θ is the incident angle) of the incident wave; however, if multiple types and directions of the incident wave exist, then c_a can be set as the lowest physical wave velocity (e.g., C_{s0}) of the studied medium as suggested by the MTF developers. The numerical simulations in the ensuing section show that these guidelines can generate satisfactory absorbing effects of MTF when applied to a saturated poroelastic medium.

7.2.2 Numerical Experiments

In order to investigate its wave-absorbing capacity in computations, the second-order MTF is implemented into a self-developed finite element code, which contains saturated soil elements based on the ***u-w*** formulation of Biot's theory. The second-order MTF (N=2) is employed here because only two steps back in time and four steps back in space grid are needed, which is easier to be implemented and is cheaper to be stored in a finite element analysis.

Five numerical experiments are carried out in this section to study transient wave propagations in the saturated poroelastic medium. In the following computations, the 2D model is spatially discretized by the four-node or nine-node quadrilateral element. The 3D model is discretized by the eight-node brick element. Here the quadrilateral elements and the brick elements are adopted because they are more suitable to forming a line, on which the MTF formulation is based, normal to the artificial boundary when compared with the triangular elements and the tetrahedral elements. The time integration is accomplished using the implicit Newmark method. Numerical damping is introduced by taking the Newmark parameter to be 0.6 to remove disturbances caused by the sudden application of external excitations. A uniform grid is adopted for the spatial discretization with the grid size $\Delta s/\lambda_{\min} \leq 0.1$. The time-step length $\Delta t/T$ is set so that the interpolation coefficient $s=c_a\Delta t/\Delta s$ keeps the value of 0.5 for different values of the artificial wave velocity c_a.

As the MTF formulation given in Eq. (7.2.6) is essentially a forward Euler scheme, we take the interpolation coefficient s=0.5 so that it meets the Courant-Friedrichs-Lewy (CFL) condition $c_a\Delta t/\Delta s \leq 1/\sqrt{2}$[84]. Here the finite element model with MTF applied to its boundaries constitutes an initial boundary value problem whose stability is a rather complicated topic, especially in view of: (1) the implicit Newmark scheme for the interior nodes is not compatible with the explicit forward Euler scheme for the boundary nodes; (2) the viscous coupling between the solid and fluid phases of the poroelastic medium renders the wave propagation dispersive; and (3) the finite element discretization and the numerical damping also make the wave propagation dispersive. Based on the GKS stability criterion[85], Higdon[86] proved that the finite-difference equation with "the space-time extrapolation" boundary condition, which corresponds to the MTF formulation in Eq. (7.2.1), is stable with the exception at zero frequency. Similarly, for the finite-difference equations, Wagner and Chew[70] demonstrated that the MTF formulation, after the quadratic space interpolation as in Eq. (7.2.6), is marginally stable at the zero frequency. The suggested remedy is to add a small negative value δ (=−0.04 s~−0.3 s) to the interpolation coefficient T_{11} (see Eq. (7.2.3)). From our numerical simulations, we note that the proposed stability criterion and the remedy of the MTF associated with finite-difference equations are conservative for the finite element scheme adopted here when the following conditions are used: (1) implicit Newmark time marching scheme; (2) square elements meshed near the artificial boundaries; and (3) CFL condition for boundary nodes.

One-Dimensional 1D Longitudinal Wave Propagation

A saturated soil column, with a width of 1 cm (x axis) and a height of 10 cm (y axis), was established and spatially discretized using 100 equally spaced four-node quadrilateral elements (Δs=0.1cm) along the y axis. The movements of both the soil-skeleton and pore-water phases in the x direction were constrained to mimic 1D dilatational wave propagation. A Heaviside step function with an amplitude of 0.001 cm was applied at the top of the column to achieve the soil-skeleton displacement u_y. The top of the column is impervious, i.e., w_y=0. At the base of the column, the MTF was applied to both the solid and fluid phases to transmit the longitudinal wave generated by the displacement input.

Material parameters of the saturated soil and velocities of the three body waves at extreme conditions are given in Table 7.2.1. The total time period was determined to include three complete reflections of the longitudinal wave for an observation point located 5 cm below the top of the column. Two values of Darcy permeability were considered, i.e., one very small (k_D=1.0×10^{-9} m/s) and one very large (k_D=0.1 m/s), to show the absorbing characteristics of MTF at high and low viscous couplings between the two phases. Figs. 7.2.6 and 7.2.7 compare the absorbing capacities of the MTF for three different values of the artificial velocity, i.e., $c_a=C_{s0}$, $c_a=C_{p0}$, and $c_a=C_{p2}$, when k_D=1.0×10^{-9} m/s and k_D=0.1 m/s, respectively. Also included in the two figures are the responses obtained for the fixed bottom of the column and the analytical solutions given by Gajo and Mongiovi[87] for the semiinfinite soil column.

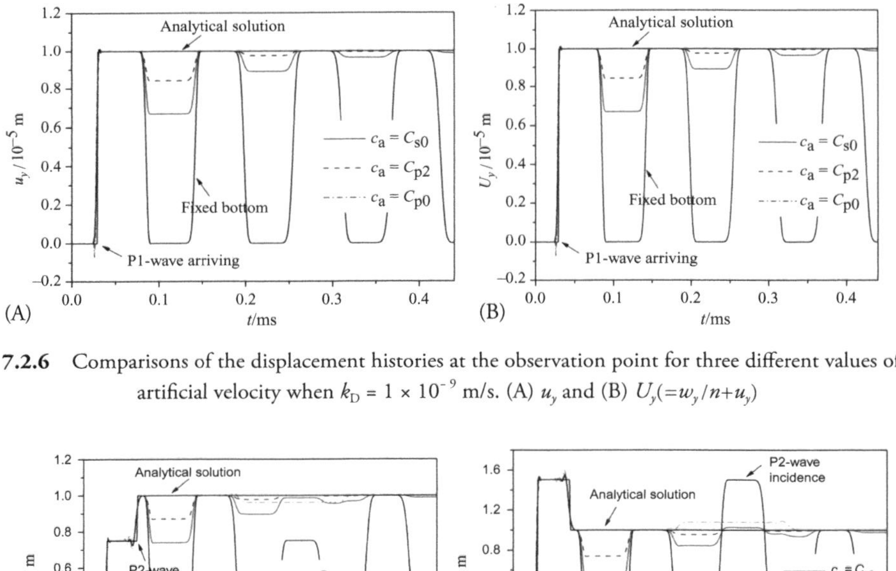

Fig. 7.2.6 Comparisons of the displacement histories at the observation point for three different values of the artificial velocity when k_D = 1 × 10^{-9} m/s. (A) u_y and (B) $U_y(=w_y/n+u_y)$

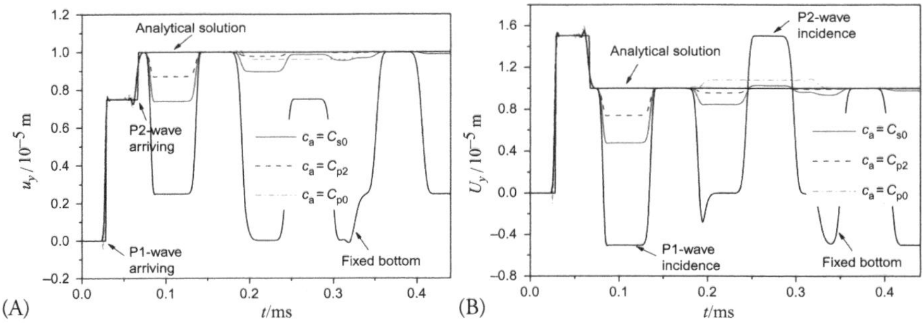

Fig. 7.2.7 Comparisons of the displacement histories at the observation point for three different values of the artificial velocity when k_D = 0.1 m/s. (A) u_y and (B) $U_y(=w_y/n+u_y)$

Only one wave front contributed by the P1 wave is shown in Fig. 7.2.6, as the high viscous coupling between the soil-skeleton and pore-fluid phase damps the P2 wave. The model configuration here generates only the longitudinal wave at normal incidence. Hence, the artificial velocity c_a should be set to C_{p0} of the longitudinal wave. As seen from Fig. 7.2.6, the line for $c_a=C_{p0}$ completely overlaps the analytical solutions with a reflection coefficient very close to zero. This agrees with the predicted value in the theoretical investigation shown in Fig. 7.2.5(a) at a normal incidence, i.e., $R_{aa}\approx0$ with $c_a=C_{p0}$ at $\theta^{I}_{1p}=0°$. When c_a decreases to C_{p2} and C_{s0}, the P1-wave reflection arises but with a decreasing reflection amplitude as the time increases, and eventually the two lines will converge to the analytical solutions. By comparing with the analytical solutions, the P1-wave reflection coefficients when $c_a=C_{p2}$ and $c_a=C_{s0}$ can be computed as 0.16 and 0.33, respectively. For the theoretical predictions, it can be seen from Fig. 7.2.5A that the P1 wave reflection coefficients R_{aa} associated with $c_a=C_{p2}$ and $c_a=C_{s0}$ are 0.17 and 0.35, respectively, at the normal incidence of the P1 wave. Thus, the reflection coefficients of the computational evaluations are slightly smaller than those of the theoretical predictions. The reason is that the theoretical predictions in Fig. 7.2.5 are carried out for $\chi=10$, i.e., the viscous coupling between the two phases of the saturated soil is low. Actually, when the viscous coupling is high ($\chi=0.1$ or 0.001), it is seen from Fig. 7.2.3A that R_{aa} associated with $c_a=C_{s0}$ decreases to 0.33 at the normal incidence, which is exactly the same as the computational result for $c_a=C_{s0}$.

When the Darcy permeability k_D increases to 0.1 m/s, the viscous coupling is very low and the P2 wave becomes a propagating wave. Therefore, two wave fronts contributed by the P1 and P2 waves are clearly visible in Fig. 7.2.7.

The lines confined in the time period from 0.08 to 0.135 ms are reflection amplitudes corresponding to the incidence of the P1 wave. The comparisons for the three different artificial velocities in this time interval are similar to those depicted in Fig. 7.2.6 but with slightly larger reflection coefficients, i.e., 0.17 with $c_a=C_{p2}$ and 0.35 with $c_a=C_{s0}$. The P1-wave reflection coefficients of the computational evaluations are the same as those of the theoretical predictions shown in Fig. 7.2.5A, which gives R_{aa}=0.17 and 0.35 when c_a equals C_{p2} and C_{s0}, respectively.

The lines confined in the time period from 0.24 to 0.3 ms are reflection amplitudes corresponding to the incidence of the P2 wave. Thus the line for $c_a=C_{p2}$ completely coincides with the analytical solutions with a reflection coefficient very close to zero, while some reflections exist for the other two artificial velocities. The reflection coefficients in this time interval can be computed as 0.05 for $c_a=C_{s0}$ and 0.16 for $c_a=C_{p0}$. These computational reflection coefficients agree well with the theoretical predictions made in Fig. 7.2.5B at a normal incidence, in which the P2 wave reflection coefficient R_{bb} equals 0.0006, 0.05, and 0.17 when c_a takes the value of C_{p2}, C_{s0}, and C_{p0}, respectively.

One-Dimensional (1D) Shear Wave Propagation

The same saturated soil column was used to study 1D shear wave propagation. The column was spatially discretized with 100 nine-node quadrilateral elements (Δs=0.1cm). The movements of both the soil-skeleton and pore-water phases along the y axis were constrained to mimic 1Dshear wave propagation. A Heaviside step function with an amplitude of 0.001 cm was applied to the soil-skeleton displacement u_x at the top of the column. An impervious boundary, i.e., $w_x=0$, was assumed for the top of the column. Other model settings and material parameters were the same as those for Figs. 7.2.6 and 7.2.7.

Figs. 7.2.8 and 7.2.9 compare the displacement histories at the observation point (5 cm below the top of the column) between the three different values of the artificial velocity, i.e., $c_a = C_{s0}$, $c_a = C_{p0}$, and $c_a = C_{p2}$, respectively, for low ($k_D = 1.0\times10^{-9}$m/s) and high ($k_D = 0.1$m/s) Darcy permeability.

The model configuration used here allows only the normal incidence of the SV wave. Thus the artificial velocity can be set to C_{s0} of the shear wave. As observed from Figs. 7.2.8 and 7.2.9, the MTF with $c_a = C_{s0}$ almost perfectly absorbs the SV wave even at its first reflection, while a certain reflection occurs for the other two artificial velocities during the three reflections. The SV-wave reflection coefficients with $c_a = C_{p2}$ and $c_a = C_{p0}$ can be computed, respectively, as 0.03 and 0.3 from Figs. 7.2.8 and 7.2.9. Again, the computational reflection coefficients agree with the theoretical predictions made in Fig. 7.2.5C, which gives $R_{cc} = 0.04$ and 0.31 for $c_a = C_{p2}$ and C_{p0}, respectively, for a normal incidence of the SV wave.

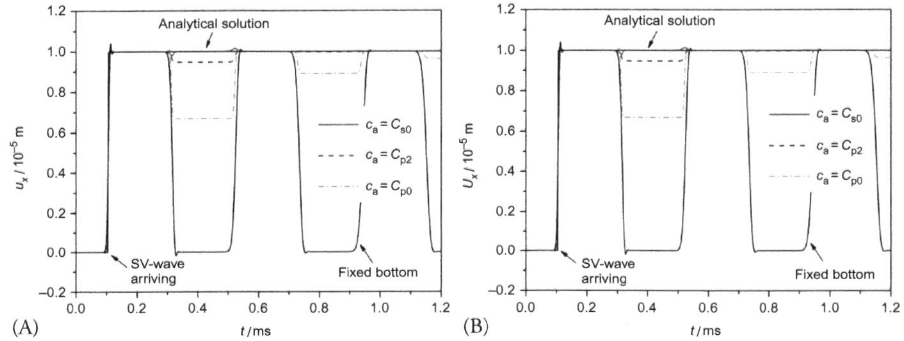

Fig. 7.2.8 Comparisons of the displacement histories at the observation point for three different values of the artificial velocity when $k_D = 1\times10^{-9}$ m/s. (A) u_x and (B) $U_x(=w_x/n+u_x)$

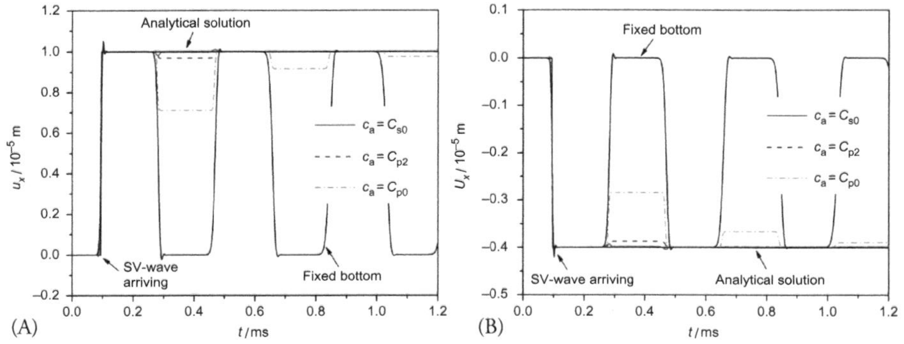

Fig. 7.2.9 Comparisons of the displacement histories at the observation point for three different values of the artificial velocity when $k_D = 0.1$ m/s. (A) u_x and (B) w_x

Plain-Strain Waves in a Saturated Square Soil Medium

A schematic view of the plain-strain square domain occupied by a saturated soil medium is shown in Fig. 7.2.10A. A uniformly distributed edge loading with a width of 2 m was applied vertically to the top left corner of the soil medium. The loading was a triangular impulse load with an amplitude of 20 kN/m and lasting for T=0.2 s, as shown in Fig. 7.2.10B.

The soil domain was spatially discretized by nine-node quadrilateral elements with a uniform size of 2 m×2 m (Δs=2 m). The left side of the square domain was a symmetric boundary (u_x=w_x=0) and the surface was impermeable (w_y=0). The right and bottom sides were applied with MTF. Material parameters of the saturated soil were taken from Li and Song[83] λ=3.184×10^8 Pa, μ=7.96×10^7 Pa, ρ_s=2.65×10^3 kg/m^3, ρ_f=1.0×10^3 kg/m^3, a_∞=1.0, n=0.286, α=0.99, M=5741.8 MPa, and k_D=1.0×10^{-13} m/s. The shear wave velocities at the fully undrained condition can be obtained as C_{s0}=191.1 m/s. As the type and the incident angle of the wave arriving at the artificial boundary were unknown, we set the artificial wave velocity to be c_a=C_{s0}. The time integration involved 400 steps of Δt=0.005 s for a total period of 2 s.

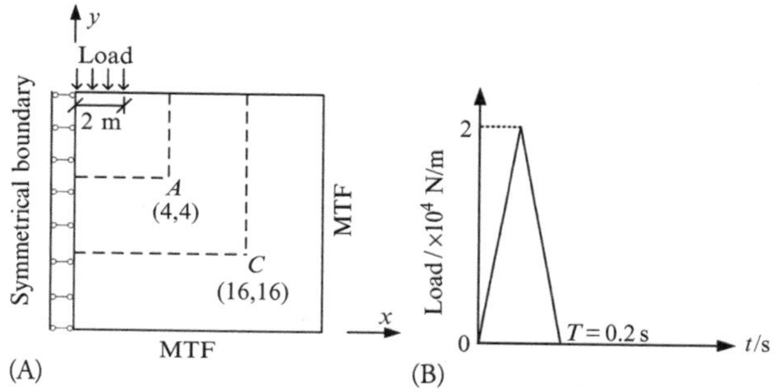

Fig. 7.2.10 Schematic view of a saturated square soil medium and loads. (A) problem geometry and (B) distributed loading

Fig. 7.2.11A and B compares the vertical soil skeleton displacement at the observation point A with a model size 20 m × 20 m and the point C with a model size 100 m × 100 m, respectively, to the reference solution that has a model size 600 m × 600 m and a fixed boundary. Also included in the two figures are solutions obtained with the viscous-spring boundary condition (ABC I) and the high-order boundary condition (ABC II), as given by Li and Song[83]. The Fourier spectrum of the triangle impulse loading was centered at zero frequency; thus the interpolation coefficient T_{11} was perturbed by adding δ=−0.04 s to avoid possible instabilities.

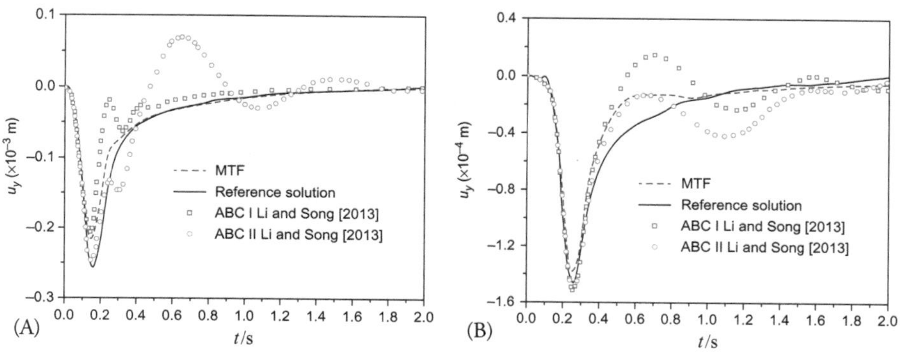

Fig. 7.2.11 Comparisons between the displacement histories obtained with MTF and the reference solutions. (A) observation point A with model size 20 m × 20 m and (B) observation point C with model size 100 m × 100 m

It can be seen from Fig. 7.2.11A that the comparison between the solution with MTF and the reference solution was generally good except for a small discrepancy in amplitude around 0.15 s. When the model size increased (Fig. 7.2.11B), better coincidence in the amplitudes of the two solutions was achieved but with some reflections occurring around =0.6s. In contrast, the solutions obtained with the two ABCs proposed by Li and Song[83] showed some reflections during the entire time period.

Waves Generated by Moving Loads

A saturated strip soil medium depicted in Fig. 7.2.12 was used to study the absorbing capacity of MTF for waves generated by a moving point load. The observation point A was located at the middle surface. The vertical point load, with an amplitude of F_0 and lasting time of T, moved at a constant speed c_0 at the surface of the soil medium. The traverse range c_0T was symmetrically distributed on both sides of A. The soil width was set to be c_0T+20 m, i.e., 10 m added at each side of the traverse range, and the soil height was 10 m. The left and right sides of the strip soil were applied with the MTF or fixed, and the bottom side was fixed.

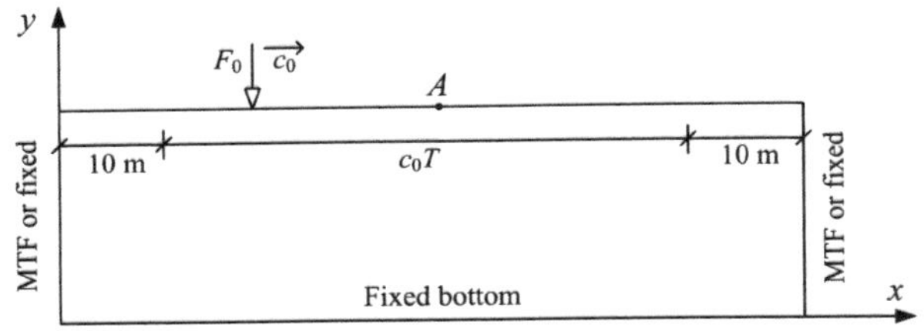

Fig. 7.2.12 Schematic view of saturated strip soil medium and the moving point load

The soil domain was spatially discretized by four-node quadrilateral elements with a uniform size of 0.5 m×0.5 m (Δs=0.5 m). Material parameters of the saturated soil were taken as $\lambda=3.0\times10^7$ Pa, $\mu=2.0\times10^7$ Pa, $\rho_s=2.7\times10^3$ kg/m^3, $\rho_f=1.0\times10^3$ kg/m^3, $a_\infty=1.0$, n=0.4, α=1.0, M=5500 MPa, and $k_D=1.0\times10^{-7}$ m/s. The shear wave velocities at the fully undrained condition can be obtained as C_{s0}=99.5 m/s. Parameters of the moving point load were F_0=1 k/N and c_0=10 m/s or 100 m/s. The artificial wave velocity c_a was set to C_{s0}. The time integration involved 1000 steps of Δt=0.002 s for the time period T=2 s.

For the observation point A, the vertical soil-skeleton and pore-fluid relative displacements are presented in Fig. 7.2.13. Comparisons have been made for responses obtained from the MTF, the fixed boundary and the extended mesh. In the extended mesh, the soil width was enlarged to 300 m with fixed left and right sides. Hence, the reflected SV waves from the fixed boundary would arrive at A later than 2.5 s, which was out of the time window (2 s) considered here. In the figure, the time t, the displacements (u_y and w_y) and the load speed c_0 were divided by T, $u_0=F_0/\mu a^2$ (a=1 m) and C_{s0}, respectively, to make them dimensionless.

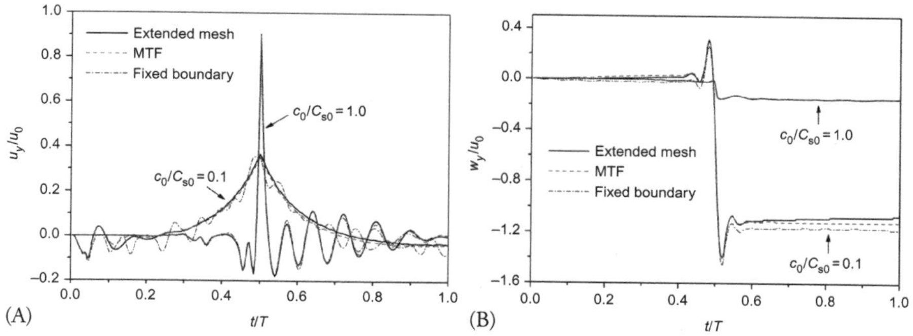

Fig. 7.2.13 Comparisons of the vertical displacement histories at the observation point when the left and right sides are fixed, applied with MTF and extended. (A) u_y and (B) $U_y(=w_y/n+u_y)$

It can be seen from Fig. 7.2.13 that the solution obtained with MTF agrees well with that of the extended mesh, while obvious reflections occur in solutions with the fixed boundaries when the load velocity is low. When the load velocity increases, the model width has to be enlarged to accommodate the increased traverse range of the moving load. Thus the solution for fixed boundaries was close to that of the extended mesh when $c_0/C_{s0}=1$. However, small discrepancies between the two solutions can still be observed in the dimensionless time range from 0.5 to 1.0, where the solution of the MTF is closer to that of the extended mesh.

Three-Dimensional (3D) Wave Propagation

A 3D cubic domain with dimensions of 20 m×20 m×20 m occupied by a saturated soil medium is shown in Fig. 7.2.14A. A point load is applied at the origin point (0 m, 0 m, 20 m) of the surface and its variation with time is shown in Fig. 7.2.14B. The observation point A was located at (5 m, 5 m, 20 m). The soil domain was spatially discretized by eight-node brick elements with a uniform size of 1m×1m×1m (Δs=1m). The x-z plane (i.e., y=0) and the y-z plane (i.e., x=0) are two symmetric planes. The bottom and the remaining two side surfaces were applied with MTF or fixed. The top surface was free and permeable.

Material parameters of the saturated soil are the same as those used in the previous section. As in the previous two numerical experiments, we set the artificial wave velocity at $c_a=C_{s0}$. The time integration involves 160 steps of Δt=0.005s for a total period of 0.8 s. Fig. 7.2.15 compares the horizontal and vertical displacement responses at the observation point to results obtained from the fixed boundary and to reference solutions obtained from an extended domain with dimensions of 45 m×45 m×45 m. It can be seen from Fig. 7.2.15 that the displacement responses obtained with MTF were very close to those from the extended mesh except for the time interval from 0.5 to 0.8 s, where some reflections from boundaries of the extended domain were observed. In contrast, significant reflections were observed for solutions obtained with the fixed boundary.

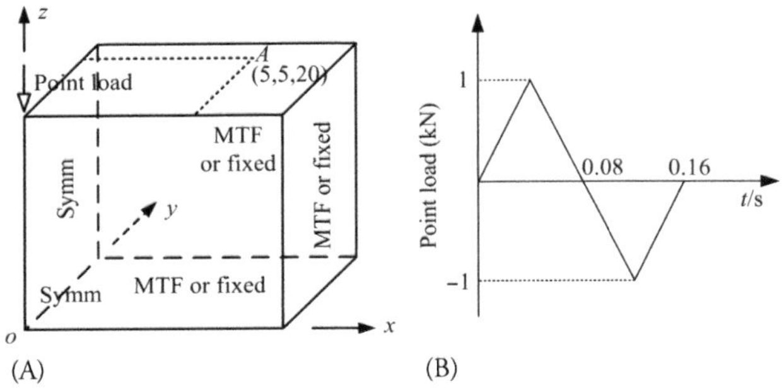

Fig. 7.2.14 Schematic view of a cubic saturated soil domain and a point load. (A) problem geometry and (B) point load

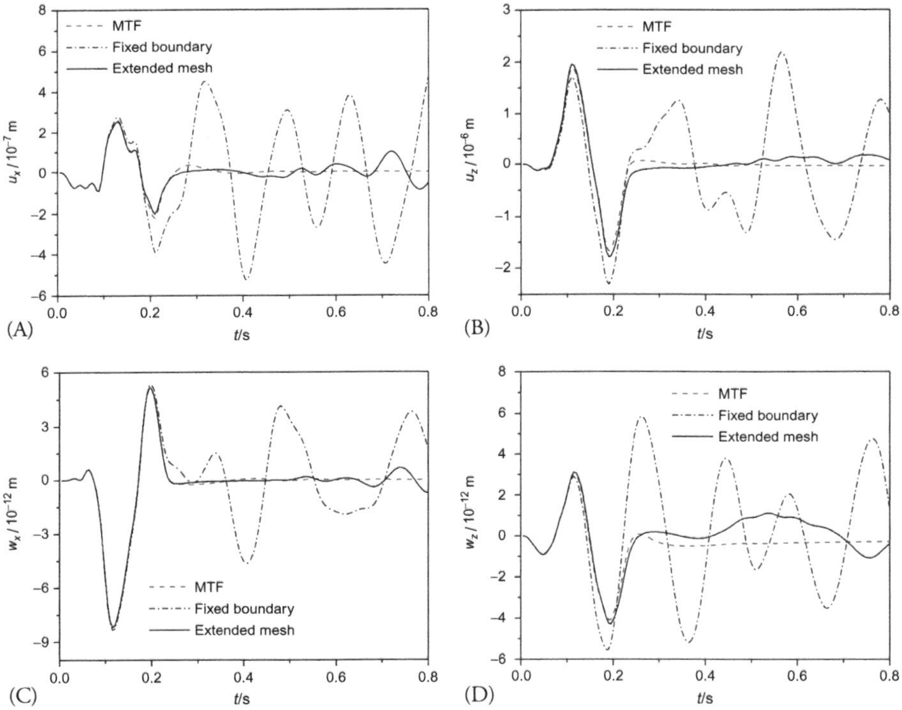

Fig. 7.2.15 Comparisons of the horizontal and vertical displacement histories at the observation point when the side and bottom surfaces are fixed, applied with MTF and extended. (A) u_x; (B) u_z; (C) w_x; and (D) w_z

7.2.3 Conclusions

In this chapter, we have developed a 3D finite element for a saturated poroelastic medium basing on the ***u-w*** formulation of Biot's theory. The accuracy of the developed element has been verified by comparing to analytical solutions and published results. As an integral part of the finite element computation, an ABC, named the MTF, has been proposed for the developed saturated soil element. Its reflection coefficients were analytically derived for the incidence of the SV, P1, and P2 waves, respectively. Parametric studies were conducted on the influence of the excitation frequency and the artificial velocity on the reflection coefficients of the MTF. Then the second-order MTF was implemented into a finite element code in order to examine its capacity in absorbing the one-dimensional longitudinal/shear wave, the plain-strain waves, the moving-load generated waves and the three-dimensional waves in the saturated poroelastic medium. It was found that the MTF is effective in absorbing waves in a 1D/2D/3D saturated poroelastic medium generated by the stationary displacement/force input and by the moving load input. Although the finite element equations of the saturated soil element were formulated in a general scheme considering the elasto-plastic behavior of the soil skeleton, only elastic behavior was considered for all the numerical experiments. It is suggested that the effectiveness of the MTF should be tested for the elasto-plastic soil in future studies.

7.3 Evaluation on Vibration Impacts of the City Railway Line S1 on the Existing Xin Z.Y. Waterworks

7.3.1 Project Background

The planned City Railway Line S1 (CRL S1 for short), striking form west to east, is located in the central area of Wenzhou City. CRL S1 passes the narrow area between the existing Jin-Wen Railway and the Xin Zhuangyuan (Xin Z.Y. for brevity) Waterworks (Fig. 7.3.1), where the railway track has been elevated using viaduct. The viaduct is very close to the Waterworks with the 12th–21th pile foundation of the elevated bridge intruding into the fence wall of the waterworks.

One of the most import functional body of the waterworks is the Flocculation-Setting Tank. The closest distance between the CRL S1 and the tank is only 9 m and the maximum distance between them is 11.3 m. Since the neighboring distance between them is too short, there are concerns that the operating of the trains on the viaduct would cause unfavorable vibrations to the Flocculation-Setting tank of the Xin Z.Y. waterworks, which would deteriorate the sealing functionality of the tank in the long run.

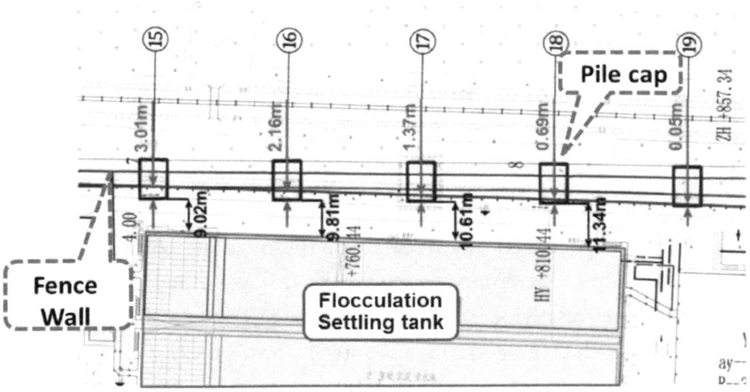

Fig. 7.3.1 Layout showing the distance between the pile foundations of the viaduct and the Flocculation-Setting tank of the Xin Z.Y. Waterworks

Thus the vibration level at the Flocculation-Setting Tank needs to be evaluated to make sure it meets the current code that sets limits on the acceptable vibrations levels for sensible infrastructures, before the actual construction of the planed CRL S1. Under this circumstance, the finite element method developed previously is used for establishing numerical 3D model including the saturated ground, the pile groups and the Flocculation-Setting Tank. Meanwhile the previously developed artificial boundary condition named Multi-transmitting Formula (MTF) is applied to the boundary faces of the model that truncate the infinite ground into a cubic body. The input to the established model will be the dynamic forces at the end of the pier which supports the bridge deck and rests on top of the pile caps. The input dynamic load originates from the interactive wheel/rail forces, which transmit all the way through the track, the bridge deck and finally to the pier.

In the following context, the analytical model for calculating the dynamic loads at the pile caps (i.e. at the end of the pier) will be described briefly. Then the vibration levels at the infrastructures of the waterworks will be predicted using a FEM model basing on the previously developed 3D saturated soil elements and the associating MTF boundary conditions.

7.3.2 Computational Models

Theoretical models have been adopted for evaluating the transferring process of the wheel/rail loads to the track structures, then to the bridge deck and supports, and finally to the pier and the pile cap. For the details of the theoretical model, one is referred to the paper by Shi and Cai et al.[88] They are omitted here for brevity.

Then the dynamic loads at the pile cap are inputted into the FEM models including the pile groups, the ground soil and the flocculation-settling tank to evaluate the vibration levels.

7.3.3 Finite Element Model on the Pile Group, the Saturated Ground and the Flocculation-Settling Tank

The Young's modulus, Poisson's ratio, mass density, length and radius of the pile are denoted as E_p、ν_p、ρ_p、l_p and d_p, respectively. Suppose there are n_{px} and n_{py} equally-spaced piles along the x and y axles of the pile cap, respectively. The spacing of the piles along the x and y axle is denoted as s_{px} and s_{py}, respectively.

Utilizing the 3D saturated soil element developed previously, the layer saturated ground, the piles and the flocculating-settling tank are discretized by 8-node brick elements. For the latter two model parts the saturated elements are reduced to single-phase elastic elements by setting small values to the parameters relating to the pore-water phase, e.g. α, M and ρ_f. For the cement-mixing piles under the tank, they are treated as the improved soils by increasing Young's modulus and the mass density.

The multi-transmitting formula developed previously is applied to the model boundaries to damp out the radiating waves, such that the reflections from the truncation boundaries would be minimized. The established FEM model including the 4 pile foundations (numbered from 15 to 18 as shown in Fig. 7.3.1) and the flocculating-settling tank is shown in Fig. 7.3.2. The distance between the pile cap and the long edge of the tank is taken as 9 m, i.e. the closest distance given in Fig. 7.3.1, to consider the most disadvantage situation.

The Newmark integration scheme has been adopted for the time marching of the dynamic analysis. The Newmark constants β=0.6, γ=0.3205 are taken such that certain amount of numerical damping is introduced to stabilize the numerical results.

For a reliable modeling on the wave propagation, the element size Δx has to be less than $\lambda_{min}/4$, where $\lambda_{min}=c_s/f_{max}$ is the minimum wavelength, c_s is the shear wave velocity and f_{max} is maximum frequency considered for the modeling.

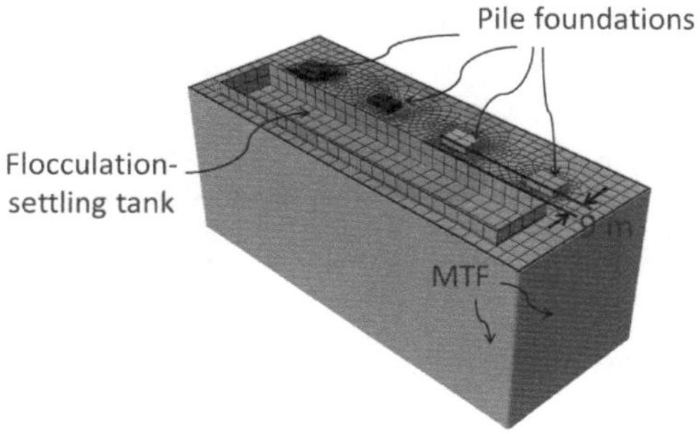

Fig. 7.3.2 FEM model including the pile foundations and the flocculation-settling tank embedded in a saturated layered ground

7.3.4 Model Parameters

According to the definitions of different variables for the train, the track, the bridge, the pier, the pile foundation, the saturated ground and the flocculation-settling tank along with its underlain cement-mixing piles, the assigned values are given in Table 7.3.1 to 7.3.7.

Table 7.3.1 Parameters for the train wheels and the rail unevenness

Parameter	Value
Wheel mass, m_w	1350 kg
Axle load, W	200 kN
Wheel/rail contact stiffness, k_H	2.8×10^9 N/m
Train speed, c	120 km/h
Number of carriages, n_v	4–6
Number of wheels per carriage, n_w	4
Wheel spacing, d_w	2.5 m
Axle spacing, l_w	15.5 m
Wavelength of rail unevenness, λ	1–10 m
Amplitude of rail unevenness, A	0.5–1 mm

Table 7.3.2 Parameters for the light-weight rubber track

Parameter	Value
Rail bending stiffness, EI_r	1.3247×10^7 N·m^2
Rail linear mass density, m_r	120 kg/m
Rail damping ratio, η_r	0.001
Rail fastener stiffness, k_{rp}	9.0×10^7 N/m
Rail fastener damping ratio, η_{rp}	0.2
Number of rail fasteners, n_{rp}	10
Rail fastener spacing, d_{rp}	0.58 m
Slab bending stiffness, EI_s	2.34×10^8 N·m^2
Slab linear mass density, m_s	2323.2 kg/m
Slab damping ratio, η_s	0.05
Number of slabs, n_s	6
Number of springs under slab, n_{sp}	4
Slab length, l_s	5.8 m
Stiffness of under-slab springs, k_{sp}	1.74×10^8 N/m
Damping ratio of under-slab springs, η_{sp}	0.25
Spacing of under-slab springs, d_{sp}	1.45 m

Table 7.3.3 Parameters for the elevated bridge

Parameter,	Value
Girder bending stiffness, EI_b	1.533×10^{11} N·m^2
Girder linear mass density, m_b	2.0163×10^4 kg/m
Girder damping ratio, η_b	0.05
Stiffness of girder support, k_{bp}	1.0×10^{13} N/m
Damping ratio of girder support, η_{bp}	0.1
Girder support spacing, d_{bp}	0.65 m
Girder length, L	34.8 m
Track spacing, $2e_y$	5.865 m

Table 7.3.4 Parameters for the pier

Parameter,	Value
Bending stiffness of pier, EI_{prx}	5.43×10^{11} N·m^2
Bending stiffness of pier, EI_{pry}	1.98×10^{11} N·m^2
Compressive stiffness of pier, EA_{pr}	3.51×10^{11} N·m^2
Pier linear mass density, m_{pr}	2.69×10^4 kg/m
Pier height, l_{pr}	13.1 m
Pier support spacing, $2e_x$	1.3 m

Table 7.3.5 Parameters for the pile cap and the piles

Parameter	Value
Edge length of pile cap, l_{cx}	10.2 m
Edge length of pile cap, l_{cy}	6.9 m
Height of pile cap, l_{cz}	2.5 m
Pile length, l_p	50 m
Pile radius, d_p	1.25 m
Pile spacing, s_{px}	3.9 m
Pile spacing, s_{py}	4.5 m
Number of piles, n_{px}	3
Number of piles, n_{py}	2
Young's modulus of pile, E_p	1.98×10^{10} N/m^2
Pile mass density, ρ_p	2400 kg/m^3
Poisson's ratio of pile, v_p	0.2

Table 7.3.6 Parameters for the layered saturated ground

Stratum No.	Name	Thickness/m	Shear wave velocity/(m/s)	Mass density/ (kg/m^3)	Poisson's ratio	Young's modulus/MPa
(2)1	Slush	18	102.07	1640	0.45	49.55
(2)2	Slush	12	130.53	1680	0.45	83.01
(4)1	Clay	10	254.12	1860	0.3	312.29
(5)1	Clay	10	318.61	1900	0.3	501.47
(5)4-1	Gravel	10	364	1950	0.3	671.76

Table 7.3.7 Parameters for the Flocculating-settling tank

Name	Length/m	Width/m	Depth/m	Wall thickness /m	Young's modulus / MPa	Poisson's ratio	Mass density/ (kg/m^3)
Flocculating-settling tank	125	21	9	0.3	3.25×10^4	0.25	2420
*Cement-mixing piles	125	21	18		600	0.3	1886

*The cement-mixing piles are treated as improved soils.

7.3.5 Results and Analysis

The dynamic forces/moments are inputted into the FEM model shown in Fig. 7.3.2 to evaluate the vibration levels at the base center of the flocculation-settling tank. Figs. 7.3.3 and 7.3.4 present the time histories of vibration velocities recorded at the center of the tank base for the rail unevenness λ=8.7 m and λ=1.16 m, respectively.

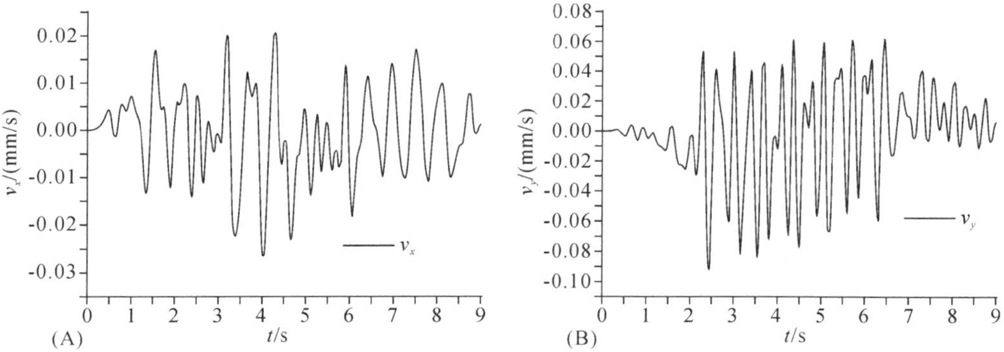

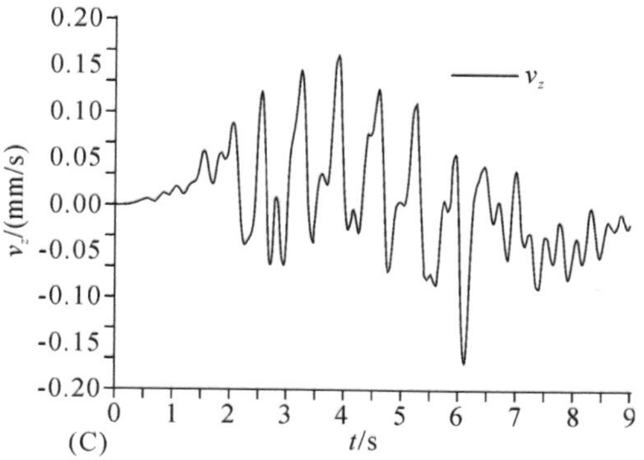

Fig. 7.3.3 Vibration velocities recorded at the center of the tank base when λ=8.7 m. (A) horizontal velocity v_x; (B) horizontal velocity v_y; and (C) vertical velocity v_z

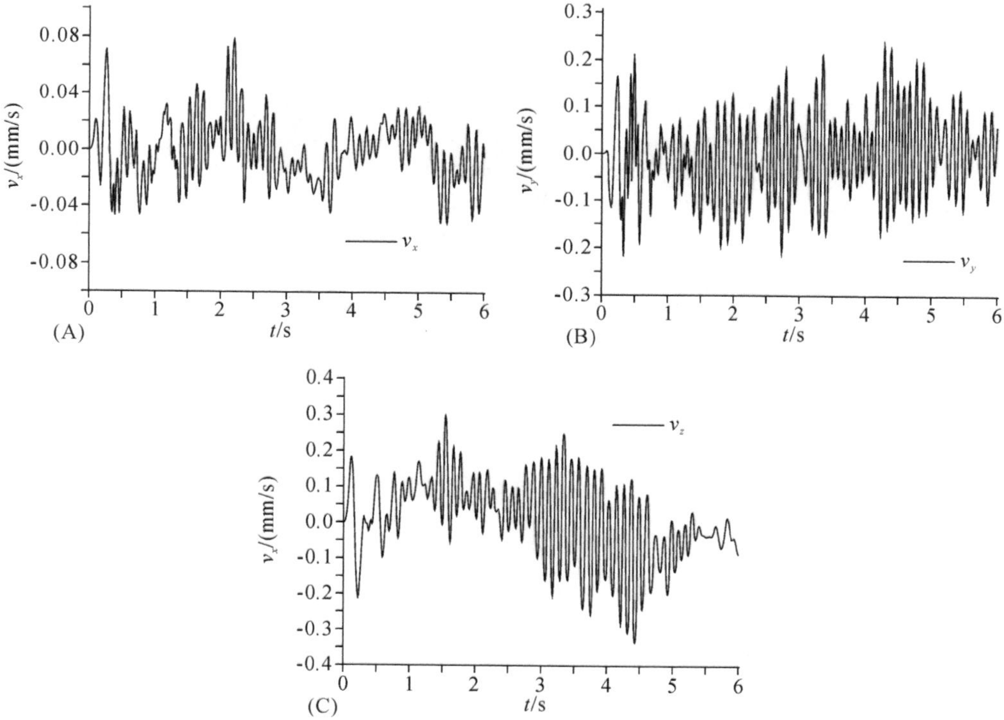

Fig. 7.3.4 Vibration velocities recorded at the center of the tank base when λ=1.16 m. (A) horizontal velocity v_x; (B) horizontal velocity v_y; and (C) vertical velocity v_z

It is within expectations that the vibration velocities are higher at λ=1.16m when compared with those at λ=8.7 m. Table 7.3.8 compares the maximum values of the vibration velocities along the three directions.

Table 7.3.8 Maximum vibration velocities at the center of the tank base

Rail unevenness	v_x/(mm/s)	v_y/(mm/s)	v_z/(mm/s)
λ=8.7 m, A=1 mm	0.026	0.093	0.173
λ=1.16 m, A=0.5 mm	0.081	0.243	0.343

It is noted that the strictest standard on acceptable vibration velocity of sensible structures is 1.0 mm/s[89]. From Table 7.3.8 it can be concluded that the maximum vibration velocities at the most disadvantage cases are within the acceptable vibration levels.

7.3.6 Conclusions

This section serves as an application of the previously developed saturated soil elements and the multi-transmitting formulas in modeling vibration and wave-propagation problems in saturated ground, in the context of the environmental vibration prediction for the operating of high-speed trains on elevated bridges.

The saturated layered ground, the pile foundations and the infrastructures of the waterworks have been modeled using the developed saturated soil element. And the multi-transmitting formulas have been applied to prevent unwanted reflections at the truncation boundaries of the FEM model.

The combination of the FEM model and the analytical model makes a successful prediction on the vibration levels of the sensible structure – the flocculation-settling tank of the Xin Z.Y. waterworks. The modeling results show that the predicted vibration levels at the most disadvantage situations are within the acceptable levels stipulated by the strictest code in China.

Appendix A

The displacement vector of the vehicle is defined as

$$\boldsymbol{Z}_{\mathrm{V}}(t)=(Z_{\mathrm{C}}(t),\ \varphi_{\mathrm{C}}(t),\ Z_{\mathrm{B1}}(t),\ \varphi_{\mathrm{B1}}(t),\ Z_{\mathrm{B2}}(t),\ \varphi_{\mathrm{B2}}(t),\ Z_{\mathrm{W1}}(t),\ Z_{\mathrm{W2}}(t),\ Z_{\mathrm{W3}}(t),\ Z_{\mathrm{W4}}(\mathrm{t})) \tag{A.1}$$

Corresponding to this displacement vector, the external load vector is defined as follows:

$$\boldsymbol{F}_{\mathrm{V}}(t)=(0,0,0,0,0,0,-P_1(t),\ -P_2(t),\ -P_3(t),\ -P_4(t))^{\mathrm{T}}=-\boldsymbol{BP}(t) \tag{A.2}$$

where

$$\boldsymbol{B}=\begin{bmatrix}\mathbf{0}_{6\times 4}\\ \boldsymbol{I}_{4\times 4}\end{bmatrix} \tag{A.3}$$

and

$$\boldsymbol{P}(t)=(P_1(t),\ P_2(t),\ P_3(t),\ P_4(t))^{\mathrm{T}} \tag{A.4}$$

is the vertical wheel-rail load vector.

The wheelset displacement vector can be written as

$$\boldsymbol{Z}_{\mathrm{W}}(t)=\boldsymbol{AZ}_{\mathrm{V}}(t) \tag{A.5}$$

where

$$\boldsymbol{A}=[\mathbf{0}_{4\times 6}\ \boldsymbol{I}_{4\times 4}]=\boldsymbol{B}^{\mathrm{T}} \tag{A.6}$$

The mass matrix is given by

$$\boldsymbol{M}_{\mathrm{V}}=\mathrm{diag}(M_{\mathrm{C}},J_{\mathrm{C}},M_{\mathrm{B}},J_{\mathrm{B}},M_{\mathrm{B}},J_{\mathrm{B}},M_{\mathrm{W}},M_{\mathrm{W}},M_{\mathrm{W}},M_{\mathrm{W}}) \tag{A.7}$$

The stiffness matrix is given by

$$\boldsymbol{K}_{\mathrm{V}}=\begin{bmatrix}
2k_2 & 0 & -k_2 & 0 & -k_2 & 0 & 0 & 0 & 0 & 0\\
0 & 2k_2 l_{\mathrm{B}}^2 & -k_2 l_{\mathrm{B}} & 0 & k_2 l_{\mathrm{B}} & 0 & 0 & 0 & 0 & 0\\
-k_2 & -k_2 l_{\mathrm{B}} & k_2+2k_1 & 0 & 0 & 0 & -k_1 & -k_1 & 0 & 0\\
0 & 0 & 0 & 2k_1 l_{\mathrm{W}}^2 & 0 & 0 & -k_1 l_{\mathrm{w}} & k_1 l_{\mathrm{w}} & 0 & 0\\
-k_2 & k_2 l_{\mathrm{B}} & 0 & 0 & k_2+2k_1 & 0 & 0 & 0 & -k_1 & -k_1\\
0 & 0, & 0 & 0 & 0 & 2k_1 l_{\mathrm{W}}^2 & 0 & 0 & -k_1 l_{\mathrm{w}} & k_1 l_{\mathrm{w}}\\
0 & 0 & -k_1 & -k_1 l_{\mathrm{w}} & 0 & 0 & k_1 & 0 & 0 & 0\\
0 & 0 & -k_1 & k_1 l_{\mathrm{w}} & 0 & 0 & 0 & k_1 & 0 & 0\\
0 & 0 & 0 & 0 & -k_1 & -k_1 l_{\mathrm{w}} & 0 & 0 & k_1 & 0\\
0 & 0 & 0 & 0 & -k_1 & k_1 l_{\mathrm{w}} & 0 & 0 & 0 & k_1
\end{bmatrix} \tag{A.8}$$

$$k_1 = \frac{k_{S1} k'_{S1} + \mathrm{i}\Omega c_{S1}\left(k_{S1} + k'_{S1}\right)}{k'_{S1} + \mathrm{i}\Omega c_{S1}} \tag{A.9}$$

The hysteretic damping can be incorporated into the suspension by introducing a complex spring stiffness.

Then, the receptance of the wheel at the wheel-rail contact point Δ_{11}^{T} is given by

$$\Delta_{11}^{\mathrm{T}} = \frac{k_1 - M_C \Omega^2}{M_C M_W \Omega^4 - k_1 M_C \Omega^2 - k_1 M_W \Omega^2} \tag{A.10}$$

Appendix B

The elements of the matrix $[\boldsymbol{U}]$ used to determine the solid displacement components of the saturated soil are:

$$
\begin{aligned}
&u_{11} = \frac{n}{r} K_n(\alpha_1 r) - \alpha_1 K_{n+1}(\alpha_1 r), \quad u_{12} = \frac{n}{r} K_n(\alpha_2 r) - \alpha_2 K_{n+1}(\alpha_2 r) \\
&u_{13} = \mathrm{i}\xi K_{n+1}(\beta r), \quad u_{14} = \frac{n}{r} K_n(\beta r) \\
&u_{21} = -\frac{n}{r} K_n(\alpha_1 r), \quad u_{22} = -\frac{n}{r} K_n(\alpha_2 r), u_{23} = \mathrm{i}\xi K_{n+1}(\beta r), u_{24} = -\frac{n}{r} K_n(\beta r) + \beta K_{n+1}(\beta r) \\
&u_{31} = \mathrm{i}\xi K_n(\alpha_1 r), \quad u_{32} = \mathrm{i}\xi K_n(\alpha_2 r), \quad u_{33} = \beta K_n(\beta r), \quad u_{34} = 0
\end{aligned}
\tag{B.1}
$$

The elements of the matrix $[\boldsymbol{W}]$ used to determine the relative fluid displacement in direction r are:

$$
\begin{aligned}
&w_{11} = \xi_{\mathrm{f}}(\frac{n}{r} K_n(\alpha_1 r) - \alpha_1 K_{n+1}(\alpha_1 r)), \quad w_{12} = \xi_{\mathrm{s}}(\frac{n}{r} K_n(\alpha_2 r) - \alpha_2 K_{n+1}(\alpha_2 r)) \\
&w_{13} = \xi_{\mathrm{t}} \mathrm{i}\xi K_{n+1}(\beta r), w_{14} = \xi_{\mathrm{t}} \frac{n}{r} K_n(\beta r)
\end{aligned}
\tag{B.2}
$$

The elements of the matrix $[\boldsymbol{T}]$ used to determine the stress components of the saturated soil are:

$$t_{11} = \left[2\mu\left(\frac{n^2 - n}{r^2} + \alpha_1^2 \right) - \eta_{\mathrm{f}} k_{\mathrm{f}}^2 \right] K_n(\alpha_1 r) + 2\mu \frac{\alpha_1}{r} K_{n+1}(\alpha_1 r)$$

$$t_{12} = \left[2\mu\left(\frac{n^2 - n}{r^2} + \alpha_2^2 \right) - \eta_{\mathrm{s}} k_{\mathrm{s}}^2 \right] K_n(\alpha_2 r) + 2\mu \frac{\alpha_2}{r} K_{n+1}(\alpha_2 r)$$

$$t_{13} = -2\mu\mathrm{i}\xi\beta K_n(\beta r) - 2\mu\mathrm{i}\xi \frac{n+1}{r} K_{n+1}(\beta r), \quad t_{14} = 2\mu \frac{n^2 - n}{r^2} K_n(\beta r) - 2\mu \frac{n}{r} \beta K_{n+1}(\beta r)$$

$$t_{21} = -2\mu \frac{n^2 - n}{r^2} K_n(\alpha_1 r) + 2\mu \frac{n}{r} \alpha_1 K_{n+1}(\alpha_1 r), \quad t_{22} = -2\mu \frac{n^2 - n}{r^2} K_n(\alpha_2 r) + 2\mu \frac{n}{r} \alpha_2 K_{n+1}(\alpha_2 r)$$

$$t_{23} = -\mu\mathrm{i}\xi\beta K_n(\beta r) - 2\mu\mathrm{i}\xi \frac{n+1}{r} K_{n+1}(\beta r), \quad t_{24} = \left(-2\mu \frac{n^2 - n}{r^2} - \mu\beta^2 \right) K_n(\beta r) - 2\mu \frac{\beta}{r} K_{n+1}(\beta r)$$

$$t_{31} = 2\mu\mathrm{i}\xi \frac{n}{r} K_n(\alpha_1 r) - 2\mu\mathrm{i}\xi\alpha_1 K_{n+1}(\alpha_1 r), t_{32} = 2\mu\mathrm{i}\xi \frac{n}{r} K_n(\alpha_2 r) - 2\mu\mathrm{i}\xi\alpha_2 K_{n+1}(\alpha_2 r)$$

$$t_{33} = \mu \frac{n}{r} \beta K_n(\beta r) - \mu\left(\xi^2 + \beta^2\right) K_{n+1}(\beta r), t_{34} = \mu\mathrm{i}\xi \frac{n}{r} K_n(\beta r)$$

$$t_{41} = \left(2\mu \frac{n - n^2}{r^2} - \eta_{\mathrm{f}} k_{\mathrm{f}}^2 \right) K_n(\alpha_1 r) - 2\mu \frac{\alpha_1}{r} K_{n+1}(\alpha_1 r)$$

$$t_{42} = \left(2\mu \frac{n - n^2}{r^2} - \eta_{\mathrm{s}} k_{\mathrm{s}}^2 \right) K_n(\alpha_2 r) - 2\mu \frac{\alpha_2}{r} K_{n+1}(\alpha_2 r)$$

$$t_{43}=2\mu\mathrm{i}\xi\frac{n+1}{r}K_{n+1}(\beta r),t_{44}=2\mu\frac{n-n^2}{r^2}K_n(\beta r)+2\mu\frac{n}{r}\beta K_{n+1}(\beta r)$$

$$t_{51}=-2\mu\mathrm{i}\xi\frac{n}{r}K_n(\alpha_1 r),\quad t_{52}=-2\mu\mathrm{i}\xi\frac{n}{r}K_n(\alpha_2 r)$$

$$t_{53}=-\mu\frac{n}{r}\beta K_n(\beta r)-\mu\xi^2K_{n+1}(\beta r),\quad t_{54}=-\mu\mathrm{i}\xi\left(\frac{n}{r}K_n(\beta r)-\beta K_{n+1}(\beta r)\right)$$

$$t_{61}=\left(-2\mu\xi^2-\eta_{\mathrm{f}}k_{\mathrm{f}}^2\right)K_n(\alpha_1 r)\quad,t_{62}=\left(-2\mu\xi^2-\eta_{\mathrm{s}}k_{\mathrm{s}}^2\right)K_n(\alpha_2 r)$$

$$t_{63}=2\mu\mathrm{i}\xi\beta K_n(\beta r),t_{64}=0 \tag{B.3}$$

where $\eta_{\mathrm{f,s}}=\lambda+\alpha M(\alpha+\xi_{\mathrm{f,s}})$. The elements of the matrix [$\boldsymbol{P}$] used to determine the pore pressure of the saturated soil are:

$$p_{11}=(\alpha+\xi_{\mathrm{f}})Mk_{\mathrm{f}}^2K_n(\alpha_1 r),\quad p_{12}=(\alpha+\xi_{\mathrm{s}})Mk_{\mathrm{s}}^2K_n(\alpha_2 r),\quad p_{13}=0,\quad p_{14}=0 \tag{B.4}$$

Coefficients of the tunnel shell

$$a_{11}=-v\mathrm{i}\boldsymbol{\xi}+\frac{h^2}{12}(\mathrm{i}\boldsymbol{\xi})^3+\frac{h^2}{12a^2}\frac{1-v}{2}\mathrm{i}\boldsymbol{\xi}n^2,\quad a_{12}=-\frac{n}{a}-\frac{h^2}{12a}\frac{3-v}{2}\boldsymbol{\xi}^2n$$

$$a_{13}=\frac{\rho_t a(1-v^2)}{E}\omega^2-\frac{h^2}{12}\left(a\boldsymbol{\xi}^4+\frac{2}{a}\boldsymbol{\xi}^2n^2+\frac{1}{a^3}n^4\right)-\frac{1}{a}+\frac{h^2}{12a^3}(2n^2-1)$$

$$a_{21}=-\frac{1+v}{2}\mathrm{i}\xi n,\quad a_{22}=\frac{\rho_t a(1-v^2)}{E}\omega^2-\frac{a(1-v)}{2}\left(1+\frac{h^2}{4a^2}\right)\xi^2-\frac{n^2}{a},\quad a_{23}=-\frac{n}{a}-\frac{h^2}{12a}\frac{3-v}{2}\xi^2n \tag{B.5}$$

$$a_{31}=\frac{\rho_t a(1-v^2)}{E}\omega^2-a\xi^2-\frac{1-v}{2a}\left(1+\frac{h^2}{12a^2}\right)n^2,\quad a_{32}=\frac{1+v}{2}\mathrm{i}\xi n$$

$$a_{33}=v\mathrm{i}\xi-\frac{h^2}{12}(\mathrm{i}\xi)^3-\frac{h^2}{12a^2}\frac{1-v}{2}\mathrm{i}\xi n^2$$

Appendix C

The variables κ_1, q, s, χ_i (i = 1, 2, 3, 4, 5), and γ_j (j = 1, 2, 3) are given by

$$\kappa_1 = \frac{n\rho^* a_0}{inb^* - \rho^* a_0} \tag{C.1}$$

$$q = \sqrt{\xi^2 + \kappa_2} \tag{C.2}$$

$$s = \sqrt{\xi^2 - a_0^2(1 + \rho^*\kappa_1)} \tag{C.3}$$

$$\chi_1 = \frac{\kappa_3}{\kappa_2}, \quad \chi_2 = \xi\frac{\lambda^* + 1 - \chi_1(1+\kappa_1)}{a_0^2 + \rho^* a_0^2\kappa_1 + \kappa_2} \tag{C.4}$$

$$\chi_3 = \frac{-\xi(1+\kappa_1)}{a_0^2(1+\rho^*\kappa_1)}, \quad \chi_4 = \chi_2\frac{q}{\xi}, \quad \chi_5 = \frac{\xi}{s} \tag{C.5}$$

$$\gamma_1 = \lambda^* - 2q\chi_4, \quad \gamma_2 = -2q\chi_2, \quad \gamma_3 = -\frac{s^2 + \xi^2}{s} \tag{C.6}$$

where

$$\kappa_2 = a_0^2\frac{\rho^* + \kappa_1(2\rho^* - 1)}{\kappa_1(\lambda^* + 2)}, \quad \kappa_3 = \rho^* a_0^2\frac{1+\kappa_1}{\kappa_1} \tag{C.7}$$

The variable d is given by

$$d = a_0\sqrt{(1+\rho^*\kappa_1)} \tag{C.8}$$

The variables q, s, and d are selected such that $\mathrm{Re}(q) > 0$, $\mathrm{Re}(s) > 0$, and $\mathrm{Re}(d) > 0$.

Appendix D

The elements $\boldsymbol{Q}_{11}$, $\boldsymbol{Q}_{12}$, $\boldsymbol{Q}_{21}$, and $\boldsymbol{Q}_{22}$ are given by

$$\boldsymbol{Q}_{11}=\begin{bmatrix}\lambda^*-2q\chi_4 & -2\xi\chi_3 & -2\xi \\ -2q\chi_2 & -2\xi\chi_3 & -\dfrac{s^2+\xi^2}{s} \\ \chi_1 & 1 & 0\end{bmatrix},\qquad \boldsymbol{Q}_{12}=\begin{bmatrix}\lambda^*-2q\chi_4 & -2\xi\chi_3 & -2\xi \\ 2q\chi_2 & 2\xi\chi_3 & \dfrac{s^2+\xi^2}{s} \\ \chi_1 & 1 & 0\end{bmatrix}$$

$$\boldsymbol{Q}_{21}=\begin{bmatrix}\chi_4 e^{-qH/r_0} & \chi_3 e^{-\xi H/r_0} & \chi_5 e^{-sH/r_0} \\ \chi_2 e^{-qH/r_0} & \chi_3 e^{-\xi H/r_0} & e^{-sH/r_0} \\ -q\chi_1 e^{-qH/r_0} & -\xi e^{-\xi H/r_0} & 0\end{bmatrix},\qquad \boldsymbol{Q}_{22}=\begin{bmatrix}-\chi_4 e^{-qH/r_0} & -\chi_3 e^{-\xi H/r_0} & -\chi_5 e^{-sH/r_0} \\ \chi_2 e^{-qH/r_0} & \chi_3 e^{-\xi H/r_0} & e^{-sH/r_0} \\ q\chi_1 e^{-qH/r_0} & \xi e^{-\xi H/r_0} & 0\end{bmatrix} \tag{D.1}$$

Appendix E

Expressions for q_1, q_2, q_3, c_{21}, c_{22}, c_{31}, c_{32}, c_{41}, c_{42}, c_{43}, and $f(\varepsilon)$ appearing previously are given by

$$q_1=\sqrt{\varepsilon^2-\frac{1}{2}\left(\beta_4+\sqrt{\beta_4^2-4\beta_5}\right)},\quad q_2=\sqrt{\varepsilon^2-\frac{1}{2}\left(\beta_4-\sqrt{\beta_4^2-4\beta_5}\right)}$$
$$q_3=\sqrt{\varepsilon^2-\omega_0^2-\rho^*\omega_0^2D_1} \tag{E.1}$$

$$c_{21}=\frac{\beta_2+\beta_1(q_1^2-\varepsilon^2)}{\beta_3},\quad c_{22}=\frac{\beta_2+\beta_1(q_2^2-\varepsilon^2)}{\beta_3},\quad c_{31}=\frac{\varepsilon(\lambda^*+1)-c_{21}\varepsilon(\alpha+D_1)}{q_1^2+\omega_0^2+\rho^*\omega_0^2D_1-\varepsilon^2}$$
$$c_{32}=\frac{\varepsilon(\lambda^*+1)-c_{22}\varepsilon(\alpha+D_1)}{q_2^2+\omega_0^2+\rho^*\omega_0^2D_1-\varepsilon^2},\quad c_{41}=\frac{\varepsilon c_{31}-1}{q_1},\quad c_{42}=\frac{\varepsilon c_{32}-1}{q_1},\quad c_{43}=\frac{\varepsilon}{q_3} \tag{E.2}$$

$$f(\varepsilon)=\frac{c_{21}c_{42}-c_{22}c_{41}+c_{43}D_2}{(\lambda-2q_2c_{42})c_{21}-(\lambda-2q_1c_{41})c_{22}-2q_3c_{43}D_2} \tag{E.3}$$

for the unsealed boundary

$$f(\varepsilon)=\frac{-q_1c_{42}+q_2c_{41}+q_2c_{43}D_3}{[(\lambda+2\varepsilon)q_2c_{21}-2\varepsilon q_2c_{31}+q_2]-[(\lambda+2\varepsilon)q_1c_{22}-2\varepsilon q_1c_{32}+q_1]-2\varepsilon q_2D_3} \tag{E.4}$$

for the sealed boundary, where

$$D_1=\frac{n\rho^*\omega_0}{inb^*-\rho^*\omega_0},\quad \beta_1=D_1(\lambda^*+2),\quad \beta_2=D_1\omega_0^2(1+D_1\rho^*)-\rho^*\omega_0^2(\alpha+D_1)^2$$
$$\beta_3=\frac{\rho^*\omega_0^2(\alpha+D_1)}{M^*},\quad \beta_4=\frac{\beta_2}{\beta_1}-\frac{\rho^*\omega_0^2}{M^*D_1},\quad \beta_5=-\frac{\rho^*\omega_0^2\beta_2}{M^*D_1\beta_1}-\frac{(\alpha+D_1)\rho^*\omega_0^2\beta_3}{D_1\beta_1} \tag{E.5}$$
$$D_2=\frac{(q_1c_{31}+\varepsilon c_{41})c_{22}-(q_2c_{32}+\varepsilon c_{42})c_{21}}{q_3+\varepsilon c_{43}},\quad D_3=\frac{q_1q_2(c_{31}-c_{32})+\varepsilon q_2c_{41}-\varepsilon q_1c_{42}}{q_1q_3+\varepsilon q_1c_{43}}$$

In these equations, the variables q_1, q_2, and q_3 are selected such that $\mathrm{Re}(q_1)>0$, $\mathrm{Re}(q_2)>0$, and $\mathrm{Re}(q_3)>0$.

Appendix F

Expressions for g_1, g_2, g_3, l_{21}, l_{22}, l_{31}, l_{32}, l_{41}, l_{42}, l_{51}, and l_{52} appearing previously are given by

$$g_1 = \sqrt{\varepsilon^2 - \beta_3}, \quad g_2 = \sqrt{\varepsilon^2 - \beta_4}, \quad g_3 = \sqrt{\varepsilon^2 - \beta_5} \tag{F.1}$$

$$\begin{aligned} &l_{21} = \frac{-M^* D_1 \beta_3 - \rho^* \omega_0^2}{\rho^* M^* \omega_0^2 (\alpha + D_1)}, \quad l_{22} = \frac{-M^* D_1 \beta_4 - \rho^* \omega_0^2}{\rho^* M^* \omega_0^2 (\alpha + D_1)}, \quad l_{31} = \frac{(\lambda^* + 1) c_{21} - (\alpha + D_1)}{\beta_3 - \beta_1} \\ &l_{32} = \frac{(\lambda^* + 1) c_{22} - (\alpha + D_1)}{\beta_3 - \beta_2}, \quad l_{41} = -\frac{2q_1}{\varepsilon}\left(c_{21} + q_1^2 c_{31}\right), \quad l_{42} = -\frac{2q_2}{\varepsilon}\left(c_{22} + q_2^2 c_{32}\right) \\ &l_{51} = \lambda c_{21} - 2q_1^2 c_{31}, \quad l_{52} = \lambda c_{22} - 2q_2^2 c_{32} \end{aligned} \tag{F.2}$$

$$D_1 = \frac{n\rho^* \omega_0}{inb^* - \rho^* \omega_0} \tag{F.3}$$

$$\begin{aligned} &\beta_1 = \frac{D_1 \omega_0^2 (1 + D_1 \rho^*) - \rho^* \omega_0^2 (\alpha + D_1)^2}{D_1 (\lambda^* + 2)} - \frac{\rho^* \omega_0^2}{M^* D_1}, \quad \beta_2 = -\frac{\rho^* \omega_0^4 (1 + D_1 \rho^*)}{M^* D_1 (\lambda^* + 2)} \\ &\beta_3 = \frac{1}{2}\left(\beta_1 + \sqrt{\beta_1^2 - 4\beta_2}\right), \quad \beta_4 = \frac{1}{2}\left(\beta_1 - \sqrt{\beta_1^2 - 4\beta_2}\right), \quad \beta_5 = \omega_0^2 + \rho^* \omega_0^2 D_1 \end{aligned} \tag{F.4}$$

In these equations, the variables g_1, g_2, and g_3 are selected such that $\mathrm{Re}(g_1) > 0$, $\mathrm{Re}(g_2) > 0$, and $\mathrm{Re}(g_3) > 0$.

Appendix G

$$u_1 = \pi a A_1 \left\{ \left[\lambda k^2\left(a_1^2-1\right) - 2\mu k^2 - b\omega\mathrm{i}(1-B_1) - \rho_2\omega^2 B_1)\right] J_1(ka) \right.$$

$$\cdot \sum_{n=0}^{\infty}\left[-(1-\delta_{l1})\sum_{j=1}^{l-1} A_n^j K_1^n\left(kd_{jl}\right) + (1-\delta_{lN})\sum_{j=l+1}^{N}(-1)^n A_n^j K_1^n\left(kd_{jl}\right)\right] \tag{G.1}$$

$$\left. + \left[\lambda k^2\left(a_1^2-1\right) - 2\mu k^2 - b\omega i(1-B_1) - \rho_2\omega^2 B_1)\right] H_1^{(2)}(ka)\cdot A_1^l \right\}$$

$$u_2 = \pi a A_2 \left\{ \left[\lambda k^2\left(a_2^2-1\right) - 2\mu k^2 - b\omega\mathrm{i}(1-B_2) - \rho_2\omega^2 B_2)\right] J_1(ka) \right.$$

$$\cdot \sum_{n=0}^{\infty}\left[-(1-\delta_{l1})\sum_{j=1}^{l-1} A_n^j K_1^n\left(kd_{jl}\right) + (1-\delta_{lN})\sum_{j=l+1}^{N}(-1)^n A_n^j K_1^n\left(kd_{jl}\right)\right] \tag{G.2}$$

$$\left. + \left[\lambda k^2\left(a_2^2-1\right) - 2\mu k^2 - b\omega i(1-B_2) - \rho_2\omega^2 B_2)\right] H_1^{(2)}(ka)\cdot A_1^l \right\}$$

$$u_3 = \pi a A_3 \left\{ 2\mu b_1 k^3 J_1(ka)\cdot \sum_{n=0}^{\infty}\left[(1-\delta_{l1})\sum_{j=1}^{l-1} A_n^j K_1^n\left(kd_{jl}\right) - (1-\delta_{lN})\cdot \sum_{j=l+1}^{N}(-1)^n A_n^j K_1^n\left(kd_{jl}\right)\right] \right.$$

$$\left. - 2\mu b_1 k^3 H_1^{(2)}(ka)\cdot A_1^l \right\} \tag{G.3}$$

v_1, v_2, v_3 = similar expressions to u_1, u_2, u_3 with A_n^j, A_1^l, and $K_1^n(kd_{jl})$ replaced by B_n^j, B_1^l, and $L_1^n(kd_{jl})$, respectively.

$$w_1 = 2\pi a\mu k a_1 A_1 \left\{ 2J_0'(ka) + \sum_{n=0}^{\infty}\left[(1-\delta_{l1})\sum_{j=1}^{l-1} A_n^j K_0^n\left(kd_{jl}\right) + (1-\delta_{lN})\sum_{j=l+1}^{N}(-1)^n A_n^j K_0^n\left(kd_{jl}\right)\right] \right.$$

$$\left. \cdot J_0'(ka) + 2H_0^{(2)'}(ka)\cdot A_0^l \right\} \tag{G.4}$$

$$w_2 = 2\pi a\mu k a_2 A_2 \left\{ 2J_0'(ka) + \sum_{n=0}^{\infty}\left[(1-\delta_{l1})\sum_{j=1}^{l-1} A_n^j K_0^n\left(kd_{jl}\right) \right.\right.$$

$$\left.\left. + (1-\delta_{lN})\sum_{j=l+1}^{N}(-1)^n A_n^j K_0^n\left(kd_{jl}\right)\right]\cdot J_0'(ka) + 2H_0^{(2)'}(ka)\cdot A_0^l \right\} \tag{G.5}$$

Appendix H

The Fourier transformations with respect to the time t and the horizontal coordinate x are defined respectively as

$$\tilde{f}(x,z,\omega)=\int_{-\infty}^{\infty} f(x,z,t)\mathrm{e}^{-\mathrm{i}\omega t}\,\mathrm{d}t$$
$$\bar{f}(\eta,z,t)=\int_{-\infty}^{\infty} f(x,z,t)\mathrm{e}^{\mathrm{i}\eta x}\,\mathrm{d}x \tag{H.1}$$

where the superscripts '~' and '–' denote a variable in the frequency and wave-number domain, respectively.

Nontrivial solutions for the eigenvalue problem result in the following dispersion relations:

$$\left[k_{\mathrm{p}}^2 C_{\mathrm{p1}}^2-\omega^2\right]\left[k_{\mathrm{p}}^2 C_{\mathrm{p2}}^2-\omega^2\right]+\mathrm{i}\frac{1}{\chi}\omega^2\left[k_{\mathrm{p}}^2 C_{\mathrm{p0}}^2-\omega^2\right]=0$$
$$\left[k_{\mathrm{s}}^2 C_{\mathrm{s}}^2-\omega^2\right]-\mathrm{i}\frac{1}{\chi}\omega^2\left[k_{\mathrm{s}}^2 C_{\mathrm{s0}}^2-\omega^2\right]=0 \tag{H.2}$$

where $\chi=\omega/\omega_0$ is the dimensionless frequency ratio; $\omega_0=b\rho/(m\rho-\rho_{\mathrm{f}}^2)$ is the characteristic frequency of the saturated poroelastic medium; C_{p1}, C_{p2}, and C_{s} are velocities of the P1, P2, and S waves, respectively, when the saturated medium is fully drained; C_{p0} and C_{s0} are velocities of the P and SV waves in the saturated medium under fully undrained conditions. These velocities under the extreme conditions (i.e., the fully drained and fully undrained conditions) are constants and given as

$$C_{\mathrm{p1}}^2+C_{\mathrm{p2}}^2=\frac{2\alpha M\rho_{\mathrm{f}}-(\lambda+2\mu+\alpha^2 M)m-\rho M}{\rho_{\mathrm{f}}^2-m\rho}\qquad C_{\mathrm{s}}^2=\frac{\mu m}{m\rho-\rho_{\mathrm{f}}^2}$$
$$C_{\mathrm{p1}}^2\cdot C_{\mathrm{p2}}^2=\frac{(\lambda+2\mu+\alpha^2 M)M-\alpha^2 M^2}{m\rho-\rho_{\mathrm{f}}^2}\qquad C_{\mathrm{s0}}^2=\frac{\mu}{\rho} \tag{H.3}$$
$$C_{\mathrm{p0}}^2=\frac{\lambda+2\mu+\alpha^2 M}{\rho}$$

From the dispersive wave numbers $k_{1\mathrm{p}}$, $k_{2\mathrm{p}}$, and k_{s} in Eq. (H.2), wave velocities and attenuation coefficients can be obtained for a given frequency ω:

$$k_{j\mathrm{p}}=\mathrm{Re}\left(k_{j\mathrm{p}}\right)+\mathrm{i}\mathrm{Im}\left(k_{j\mathrm{p}}\right)=\omega/V_{j\mathrm{p}}+\mathrm{i}q_{j\mathrm{p}}\quad(j=1,2)$$
$$k_{\mathrm{s}}=\mathrm{Re}(k_{\mathrm{s}})+\mathrm{i}\mathrm{Im}(k_{\mathrm{s}})=\omega/V_{\mathrm{s}}+\mathrm{i}q_{\mathrm{s}} \tag{H.4}$$

where $V_{j\mathrm{p}}$ (j=1,2) and V_{s} are velocities of the longitudinal (P1 and P2) and shear waves, respectively; $q_{j\mathrm{p}}$ (j=1,2) and q_{s} are attenuation coefficients of the longitudinal (P1 and P2) and shear waves, respectively. Here the symbol V that is used to represent the frequency-dependent velocities is adopted to make a distinction between the symbol C that is used to denote the frequency-independent velocities under the extreme conditions.

Substituting Eq. (H.2) back into the established eigenvalue problem, the following equation can be obtained between potential amplitudes of the solid-skeleton and the pore-fluid phases:

$$\begin{aligned} &P_j^{\mathrm{f}} = co_{j\mathrm{p}} P_j^{\mathrm{s}} \quad (j=1,2) \\ &S^{\mathrm{f}} = co_{\mathrm{s}} S^{\mathrm{s}} \end{aligned} \tag{H.5}$$

where

$$\begin{aligned} &co_{j\mathrm{p}} = \frac{-\left(\alpha M k_{j\mathrm{p}}^2 - \omega^2 \rho_{\mathrm{f}}\right)}{M k_{j\mathrm{p}}^2 - \omega^2 m + \mathrm{i}\omega b} \quad (j=1,2) \\ &co_{\mathrm{s}} = \frac{\omega^2 \rho_{\mathrm{f}}}{\mathrm{i}\omega b - \omega^2 m} \end{aligned} \tag{H.6}$$

References

[1] Biot MA. Theory of propagation of elastic waves in a fluid-saturated porous solid. 1. Low-frequency range. J Acoust Soc Am 1956; 28: 168–178.

[2] Zienkiewicz OC, Chan AHC, Pastor M, Schrefler BA, Shiomi T. Computational geomechanics with special reference to earthquake engineering. West Sussex: John Wiley & Sons Ltd; 1997: 335–355.

[3] Biot MA. Mechanics of deformation and acoustic propagation in porous media. J Appl Phys 1962; 33: 1482–1498.

[4] Cai YQ, Cao ZG, Sun HL, Xu CJ. Dynamic response of pavement on poroelastic half-space soil medium to a moving traffic load. Comput Geotech 2009; 36: 52–60.

[5] Kim SM, Roesset JM. Moving loads on a plate on elastic foundation. J Eng Mech 1998; 124(9): 1010–1016.

[6] Richart FE. Vibrations of soils and foundations. Englewood Cliffs, NJ: Prentice-Hall; 1970.

[7] Cao ZG, Cai YQ, Sun HL, Xu CJ. Dynamic responses of a poroelastic half-space from moving trains caused by vertical track irregularities. Int J Numer Anal Methods Geomech 2011; 35: 761–786.

[8] Picoux B, Le Houedec D. Diagnosis and prediction of vibration from railway trains. Soil Dyn Earthq Eng 2005; 25: 905–921.

[9] Knothe K, Grassie SL. Modelling of railway track and vehicle track interaction at high frequencies. Veh Syst Dyn 1993; 22: 209–262.

[10] Grassie SL, Gregory RW, Harrison D, Johnson KL. The dynamic response of railway track to high frequency vertical excitation. J Mech Eng Sci 1982; 24(2): 77–90.

[11] Suiker ASJ, Chang CS, Borst RD, Esveld C. Surface waves in a stratified half space with enhanced continuum properties. Part 1. Formulation of the boundary value problem. Eur J Mech A: Solids 1999; 18: 749–768.

[12] Sheng X, Jones CJC. Ground vibration generated by a harmonic load acting on a railway track. J Sound Vib 1999; 225(1): 3–28.

[13] Bian X. In-situ testing report of the ground-bourne vibrations induced by Beijing-Tianjin intercity railway lines. Institute of Geotechnical Engineering, Zhejiang University, 2008 [In Chinese].

[14] Song HP, Bian XC, Jiang JQ, Chen YM. Correlation between subgrade settlement of high-speed railroad and train operation speed. J Vib Shock 2012; 31(10): 134–140 [In Chinese].

[15] Forrest J A, Hunt H E M. A three-dimensional tunnel model for calculation of train-induced ground vibration[J]. Journal of Sound and Vibration, 2006; 294(4): 678–705.

[16] Forrest J A, Hunt H E M. Ground vibration generated by trains in underground tunnels[J]. Journal of Sound and Vibration, 2006; 294(4): 706–736.

[17] Hussein M F M, Hunt H E M. A numerical model for calculating vibration due to a harmonic moving load on a floating-slab track with discontinuous slabs in an underground railway tunnel[J]. Journal of Sound and Vibration, 2009; 321(1): 363–374.

[18] Lu J F, Jeng D S. Dynamic response of a circular tunnel embedded in a saturated poroelastic medium due to a moving load[J]. Journal of vibration and acoustics, 2006; 128(6): 750–756.

[19] Hasheminejad S M, Komeili M. Effect of imperfect bonding on axisymmetric elastodynamic response of a lined circular tunnel in poroelastic soil due to a moving ring load[J]. International Journal of Solids and Structures, 2009; 46(2): 398–411.

[20] Yuan Z, Cai Y, Cao Z. An analytical model for vibration prediction of a tunnel embedded in a saturated full-space to a harmonic point load[J]. Soil Dynamics and Earthquake Engineering, 2016; 86: 25–40.

[21] Yuan Z. Environmental vibrations induced by underground railways in the saturated soil[Dissertation]. Hangzhou: Zhejiang University, 2016 [In Chinese].

[22] Song C, Chen L, Zhen J, Liu Y, Qian C. Influence of the vibration mitigation measures in underground railways on the wall of the Bell Tower in Xi'an [J]. Journal of Disaster Prevention and Mitigation, 2008; 28 (47): 147–150 [In Chinese].

[23] Baranov VA. On the calculation of excited vibrations of embedded foundation. Polytechnical Institute of Riga: Voprosy Dynamiki: Prochnocti, N0.14; 1967: 195–207 [in Russian].

[24] Novak M, Beredugo Y. Vertical vibration of embedded footings. J Soil Mech Found Div—ASCE 1972; 98 (SM12): 1291–1310.

[25] Novak M, Beredugo Y. Coupled horizontal and rocking vibration of embedded footings. Can Geotech J 1972; 9: 477–493.

[26] Novak M, Sachs K. Torsional and coupled vibrations of embedded footings. Earthq Eng Struct Dyn 1973; 2: 11–33.

[27] Sneddon I. The use of integral transforms. New York: McGraw-Hill; 1970.

[28] Nobel B. The solution of Bessel-function dual integral equations by a multiplying-factor method. Math Proc Camb Philos Soc 1963; 59: 351–362.

[29] Lysmer J, Richart FE. Dynamic response of footings to vertical loading. J Soil Mech Found Div—ASCE 1966; 92(SM1): 65–91.

[30] Lysmer J, Kuhlemeyer RL. Finite dynamic model for infinite media. J Eng Mech Div—ASCE 1969; 95 (EM4): 859–877.

[31] Apsel RJ, Luco JE. Impedance functions for foundations embedded in a layered medium: an integral equation approach. Earthq Eng Struct Dyn 1987; 15: 213–231.

[32] Zeng X, Rajapakse RKND. Vertical vibrations of a rigid disk embedded in a poroelastic medium. Int J Numer Anal Methods Geomech 1999; 23: 2075–2095.

[33] Rajapakse RKND, Senjuntichai T. Dynamic response of multi-layered poroelastic medium. Earthq Eng Struct Dyn 1995; 24(5): 703–722.

[34] Lin CH, Lee VW, Trifunac MD. The reflection of plane waves in a poroelastic half-space saturated with inviscid fluid. Soil Dyn Earthq Eng 2005; 25: 205–223.

[35] Pao YH, Chao CM. Diffraction of elastic waves and dynamic stress concentrations. New York: Crane, Russak; 1973.

[36] Cai YQ, Ding GY, Xu CJ, Wang J. Vertical amplitude reduction of Rayleigh waves by a row of piles in a poroelastic half-space. Int J Numer Anal Methods Geomech 2009; 33: 1799–1921.

[37] Cai Y, Hu X. Vertical vibrations of a rigid embedded foundation embedded in a poroelastic half-

space. J Eng Mech 2010; 136(3): 390–398.
[38] Huang W, Wang YJ, Rokhin SI. Obliquely scattering of an elastic wave from a multilayered cylinder in a solid. Transfer matrix approach. J Acoust Soc Am 1996; 99: 2742–2754.
[39] Nobel B. The solution of Bessel-function dual integral equations by a multiplying-factor method. Math Proc Camb Philos Soc 1963; 59: 351–362.
[40] Richartr PE, Woods RD. Vibrations of soils and foundations. Englewood Cliffs, NJ: Prentice-Hall; 1970.
[41] Novak M, Beredugo Y. Vertical vibration of embedded footings. J Soil Mech Found Div—ASCE 1972; 98: 1291–1310.
[42] Novak M, Beredugo Y. Coupled Horizontal and Rocking Vibration of Embedded Footings. Can Geotech J 1972; 9: 477–493.
[43] Novak M, Sachs K. Torsional and coupled vibrations of embedded footings. Earthq Eng Struct Dyn 1973; 2: 11–33.
[44] Aviles J, Sánchez-Sesma FJ. Piles as barriers for elastic waves. J Geotech Eng 1983; 109(9): 1133–1146.
[45] Tsai PH, Feng ZY, Jen TL. Three-dimensional analysis of the isolation effectiveness of hollow pile barriers for foundation-induced vertical vibration. Comput Geotech 2008; 35(3): 489–499.
[46] Prevost JH. Wave propagation in fluid-saturated porous media: an efficient finite element procedure. Soil Dyn Earthq Eng 1985; 4(4): 183–202.
[47] Elgamal A, Yang Z, Parra E. Computational modeling of cyclic mobility and post-liquefaction site response. Soil Dyn Earthq Eng 2002; 22: 259–271.
[48] Elgamal A, Yang Z, Parra E. Modeling of cyclic mobility in saturated cohesionless soils. Int J Plast 2003; 19 (6): 883–905.
[49] Zienkiewicz OC, Chang CT, Bettess P. Drained, undrained, consolidating and dynamic behavior assumptions in soils. Geotechnique 1980; 30(4): 385–395.
[50] Gajo A, Saetta A, Vitaliani R. Evaluation of three- and two-field finite element methods for the dynamic response of saturated soil. Int J Numer Methods Eng 1994; 37: 1231–1247.
[51] Gajo A. Influence of viscous coupling in propagation of elastic waves in saturated soil. J Geotech Eng 1995; 121 (9): 636–644.
[52] Zhang H. Dynamic finite element analysis for interaction between a structure and two-phase saturated soil foundation. Comput Struct 1995; 56: 49–58.
[53] Jeremic B, Cheng Z, Taiebat M, Dafalias Y. Numerical simulation of fully saturated porous materials. Int J Numer Anal Methods Geomech 2008; 32: 1635–1660.
[54] Ketti P, Engstrom G, Wiberg NE. Coupled hydro-mechanical wave propagation in road structures. Comput Struct 2005; 32: 1719–1729.
[55] Ketti P, Lenhof B, Runesson K, Wiberg NE. Coupled simulation of wave propagation and water flow in soil induced by high-speed trains. Int J Numer Methods Geomech 2008; 32: 1311–1319.
[56] Yang JS, Jin B. Dynamic finite element method for poroelastic foundation subjected to moving loads. Chinese Quart Mech 2009; 30(1): 101–108.
[57] Gao GY, Chen QS, He JF, Liu F. Investigation of ground vibration due to trains moving on saturated multi- layered ground by 2.5D finite element method. Soil Dyn Earthq Eng 2012; 40:

87–98.
[58] Ju SH, Wang YM. Time-dependent absorbing boundary conditions for elastic wave propagation. Int J Numer Methods Eng 2001; 50: 2159–2174.
[59] Lysmer J, Kuhlemeyer RL. Finite dynamic model for infinite media. J Eng Mech Div—ASCE 1969; 95: 859–877.
[60] Deeks AJ, Randolph MF. Axisymmetric time-domain transmitting boundaries. J Eng Mech—ASCE 1994; 95: 859–877.
[61] Engquist B, Majda A. Absorbing boundary conditions for the numerical simulation of waves. Proc Natl Acad Sci U S A 1977; 74(5): 1765–1766.
[62] Clayton R, Engquist B. Absorbing boundary conditions for acoustic and elastic wave equations. Bull Seismol Soc Am 1977; 67: 1529–1540.
[63] Smith WD. A non-reflecting plane boundary for wave propagation problems. J Comput Phys 1974; 15: 492–503.
[64] Higdon RL. Radiation boundary conditions for elastic wave propagation. SIAM J Numer Anal 1990; 27: 831–870.
[65] Givoli D, Neta B. High-order non-reflecting boundary scheme for time-dependent waves. J Comput Phys 2003; 186: 24–46.
[66] Bettess P, Zienkiewicz OC. Diffusion and refraction of surface waves using finite and infinite elements. Int J Numer Methods Eng 1977; 11: 1271–1290.
[67] Liao ZP, Wong HL. A transmitting boundary for the numerical simulation of elastic wave propagation. Soil Dyn Earthq Eng 1984; 3: 133–144.
[68] Peng CB, Toksoz MN. An optimal absorbing boundary condition for finite difference modeling of acoustic and elastic wave propagation. J Acoust Soc Am 1995; 2: 733–745.
[69] Liao ZP. Extrapolation non-reflecting boundary conditions. Wave Motion 1996; 24: 117–138.
[70] Wagner RL, Chew WC. An analysis of Liao's absorbing boundary condition. J Electromagnet Waves Appl 1995; 9(7–8): 993–1009.
[71] Liao ZP. Introduction to wave motion theories for engineering. Beijing: Science Press; 2003 [in Chinese].
[72] Zienkiewicz OC. The finite element method. 3rd ed. London: McGraw-Hill; 1977.
[73] Degrande G, De Roeck G. An absorbing boundary condition for wave propagation in saturated poroelastic media. Part I. Formulation and efficiency evaluation. Soil Dyn Earthq Eng 1993; 12: 411–421.
[74] Akiyoushi T, Fuchida K, Fang HL. Absorbing boundary conditions for dynamic analysis of fluid-saturated porous media. Soil Dyn Earthq Eng 1994; 13: 387–397.
[75] Modaressi H, Benzenati I. Paraxial approximation for poroelastic media. Soil Dyn Earthq Eng 1994; 13: 117–129.
[76] Li P, Song EX. A viscous-spring transmitting boundary for cylindrical wave propagation in saturated poroelastic media. Soil Dyn Earthq Eng 2014; 65: 269–283.
[77] Nenning M, Schanz M. Infinite element in a poroelastodynamic FEM. Int J Numer Anal Methods Geomech 2011; 35: 1774–1800.
[78] Zhang H. Dynamic finite element analysis for interaction between a structure and two-phase

saturated soil foundation. Comput Struct 1995; 56: 49–58.

[79] Gajo A, Saetta A, Vitaliani R. Silent boundary conditions for wave propagation in saturated porous media. Int J Numer Methods Eng 1996; 20: 253–273.

[80] Zeng YQ, He JQ, Liu QH. The application of the perfectly matched layer in numerical modeling of wave propagation in poroelastic media. Geophysics 2001; 66(4): 1258–1266.

[81] Berenger JP. A perfectly matched layer for the absorption of electromagnetic waves. J Comput Phys 1994; 114: 185–200.

[82] Li P, Song EX. A high-order time-domain transmitting boundary for cylindrical wave propagation problems in unbounded saturated poroelastic media. Soil Dyn Earthq Eng 2013; 48: 48–62.

[83] Gajo A, Saetta A, Vitaliani R. Evaluation of three- and two-field finite element methods for the dynamic response of saturated soil. Int J Numer Methods Eng 1994; 37: 1231–1247.

[84] Chew WC. Waves and fields in inhomogeneous media. New York: Van Nostrand Reinhold; 1990.

[85] Gustafsson B, Kreiss HO, Sundstrom A. Stability theory of difference approximation for mixed initial boundary value problems. II. Math Comput 1972; 26: 649–686.

[86] Higdon RL. Absorbing boundary conditions for difference approximations to the multi-dimensional wave equation. Math Comput 1986; 47(176): 437–459.

[87] Gajo A, Mongiovi L. An analytical solution for the transient response of saturated linear elastic porous media. Int J Numer Anal Methods Geomech 1995; 19: 399–413.

[88] Shi L., Cai Y.Q., Wang P., Sun H.L. A theoretical investigation on influences of slab tracks on vertical dynamic responses of railway viaducts[J]. Journal of Sound and Vibration, 2016; 374: 138–154.

[89] Ministry of Housing and Urban-Rural Development of the People's Republic of China. GB50868-2013 Standard for allowable vibration of building engineering[S]. Beijing: China Planning Press, 2013.

Index

图书在版编目（CIP）数据

Biot 多孔弹性介质理论在关键工程领域的求解及应用 =Solutions for Biot's Poroelastic Theory in Key Engineering Fields: Theory and Applications：英文 / 蔡袁强，孙宏磊著 .—杭州：浙江大学出版社，2020.12

ISBN 978-7-308-20918-2

Ⅰ．① B… Ⅱ．①蔡… ②孙… Ⅲ．①多孔介质—土动力学—英文 Ⅳ．① TU435

中国版本图书馆 CIP 数据核字 (2020) 第 252368 号

Biot 多孔弹性介质理论在关键工程领域的求解及应用

蔡袁强　孙宏磊　著

策　　划　许佳颖
责任编辑　金佩雯　陶　杭
责任校对　贾晓燕
封面设计　周　灵
出版发行　浙江大学出版社
（杭州市天目山路 148 号　邮政编码 310007）
（网址：http://www.zjupress.com）
排　　版　杭州青翊图文设计有限公司
印　　刷　广东虎彩云印刷有限公司绍兴分公司
开　　本　787mm×1092mm　1/16
印　　张　12.75
字　　数　432 千
版 印 次　2020 年 12 月第 1 版　2020 年 12 月第 1 次印刷
书　　号　ISBN 978-7-308-20918-2
定　　价　98.00 元

浙江大学出版社市场运营中心联系方式：0571-88925591；http://zjdxcbs.tmall.com